THE MISSING LINK

THE MISSING LINK

WEST EUROPEAN NEUTRALS AND
REGIONAL SECURITY

Edited by
RICHARD E. BISSELL AND CURT GASTEYGER

Duke University Press   Durham and London   1990

© 1990 Duke University Press
All rights reserved
Printed in the United States of America
on acid-free paper ∞

Library of Congress Cataloging-in-Publication Data

The Missing link : West European neutrals and regional security/
edited by Richard E. Bissell and Curt Gasteyger.
p.   cm.—(Duke Press policy studies)
ISBN 0-8223-0953-X
1. Europe—National security. 2. Europe—Neutrality. 3. North
Atlantic Treaty Organization. I. Bissell, Richard E.
II. Gasteyger, Curt Walter, 1929-   . III. Series.
UA646.M66   1990
355'.033'04—dc20   89-17010

CONTENTS

PREFACE

This book is the result of the work of many people over several years, their separate efforts dedicated to the proposition that all European countries, in one way or another, contribute to the state of European security. It is, in many respects, simply the final report of an extensive process of consultation that involved the writing of chapters, a review conference in Geneva involving both authors and other experts in the field, and further revisions before final publication.

We are particularly grateful to those who provided intellectual and financial contributions at various stages of the project. They include, in addition to the authors of the chapters, Dr. Wilfried Aichinger, Mr. Andren Krister, Lt. Gen. Gerard Berkhof, Dr. Falk Bomsdorf, Dr. Yves Boyer, Major General Gustav Daeniker, Dr. Ingemar Dorfer, Mr. Juha Harjula, Mr. Olli Kivinen, Professor Kari Mottola, Professor Neuhold Hanspeter, Colonel Fred Schreier, Mr. Anders Sjaastad, Professor Kurt Spillmann, Mr. Bengt Sendelius, Professor W. Scott Thompson, Dr. Johan Tunberger, Dr. Pieter Volten, Dr. Nils H. Wessell, Dr. Gerhard Wettig, and Dr. John Yurechko. Special appreciation must be expressed to Dr. Theodor Winkler of the Swiss Ministry of Defense and Colonel Gerald Bouchoux of the U.S. Department of Defense.

We also wish to express our gratitude to our supportive staffs at, respectively, the Graduate Institute of International Studies in Geneva and the Center for Strategic and International Studies in Washington, D.C. Finally, we greatly appreciate the care and responsiveness of Richard Rowson, his anonymous reviewers who provided very helpful suggestions, and his staff at Duke University Press in bringing the final product to fruition.

Richard E. Bissell
Washington, D.C.
Curt Gasteyger
Geneva

INTRODUCTION

The subject of the security policies of European neutrals has to be approached from several perspectives. An American audience generally misunderstands the role of neutral countries in European politics. America's experience with implementing its own policy of neutrality during the twentieth century has been such a failure—making it seem as though neutrality acts lead to war—that such policy has little credibility with the broad American elite. A European audience, on the other hand, views neutrality in foreign policy process terms—specifically, as the contribution of the neutral countries to the CSCE (Conference on Security and Economic Cooperation in Europe) process. The forty years since armed conflict ravaged Europe have made the military considerations attached to the presence of neutral states less prominent. Both American and European audiences thus can benefit by greater exposure to issues of European neutrality.

Judgments as to European neutrals' value and reliability in facing a common security threat have varied greatly. From the American perspective, the concept of neutrality has not been proven useful for peace and security. The contribution of European neutrals to European security has not been made evident, to American eyes, in the postwar period. For neutral countries neutrality is a statement of independence, of a commitment to national security. But for Americans, neutrality can be a pejorative term, an expression of other nations' unwillingness to commit themselves to what is considered a clear political or moral choice. To be neutral in a war was thus considered morally questionable by the American public. After the war, neutrals were sometimes seen as either ambivalent in their attitudes toward the East-West conflict or free riders on the stability and defense sustained by the alliance. There has been skepticism, indeed, both in the West and in the Soviet Union, about the utility and desirability of neutral states in the era of seemingly endless confrontation.

Europeans, on the other hand, when formulating their security arrangements have taken into account the existence of neutral states for many centuries. The modern role of neutrals in European security has been of special historical interest, owing to the strategic locations of Finland, Sweden, Austria, and Switzerland. The neutrality of Ireland has not been a major issue, except during World War II, since it is far from the East-West border that has prominently defined European security issues since 1945. Yugoslavia, while taking on many characteristics of a traditional neutral, is a more recent entrant in this grouping, and is not yet equivalent to the four core neutrals. While each of these countries defines neutrality to meet its needs, overwhelmingly supported by its own public, it must be said that their neutrality has only been accepted, not clearly understood, by the military alliances to the east and the west. This book is intended not only to improve that understanding, but also to examine the ways in which security developments over the last decade have materially affected the character of neutrality.

We will not dwell on the role of the neutrals in the political processes of Western Europe. The negotiations that include the neutrals, such as the CSCE, have given some prominence to the diplomatic role of that small bloc, which has acquired a profile of its own. The neutrals are developing a habit of mutual consultation on issues on which the two alliances are unable to agree. In general, other European governments' perceptions of neutrals, given their growing diplomatic effectiveness, is changing. But many studies have been and are being written on that subject.

What has been less studied is the realignment of defense issues in Europe and its effect on the neutrals. At the nuclear level the establishment of parity or superiority by the Soviet Union in its relations with the United States has been unsettling. The use of intimidation (e.g., threats to the FRG over the installation of Pershing IIs, Soviet submarines in Swedish waters, the "offer" by the Soviet Union to promise no nuclear attacks against countries without nuclear weapons) to achieve political ends has increasingly concerned the neutrals. The focus on economic assets—the East-West natural gas pipeline, strategic trade in high technology items—as part of the military balance has increased pressure on the neutrals.

The growth of neutralist sentiment in NATO countries, especially among the successor generation, has led NATO observers to examine the effects of neutrality on security in the alliance countries. The perceptions of neutral issues among NATO states in the last ten years have changed. For instance, there is less concern about "Finlandization" and more about "Hollandization." Neutrality's constituency has clearly grown within the young generation in some NATO countries. And, indeed, policies of the European states, whether allied with one side or the other, or neutral, have served to blur the cold war security distinctions in Europe. A trend toward some kind of neutrality in other European countries could discredit neutrality as practiced by the five countries in focus in this volume, as well as overburden the amount of neutrality Europe can bear. In other words, how many neutral states can the European international system absorb without major repercussions for European security and stability?

As the contributions to this volume illustrate, neutrality has been incorporated into the policies and societies of Finland, Switzerland, Sweden, Yugoslavia, and Austria in many ways and at many levels. For Switzerland there is a long and well-established legal basis for neutrality, which is validated by centuries of interaction with neighbors that have chosen to respect it. For countries like Austria the legal basis is far more recent and thus not yet fully tested as a cornerstone of state policy.

There are similarities as well as differences among the neutrals under study. In physical size and scale of economies we are talking about the "middle powers" of Europe. Finland and Sweden are the larger countries (at 337,000 and 449,000 square kilometers respectively), whereas Switzerland, Sweden, and Yugoslavia are at the high end of the GNP scale. How those basics are converted into defense strength, however, depends on other factors. For instance, an examination of recent years indicates that about 4.8 percent of Yugoslavia's central budget goes into defense, whereas Sweden has a comparable 6.5 percent figure. Austria, with its 3.9 percent figure, suffers from both a smaller base as well as proportion devoted to defense. In terms of effective financial muscle, then, there is considerable disparity between, say, Sweden and Finland—by a proportion of 3:1.

A quite different issue comes to the fore when examining military manpower. Not only do the various neutrals begin with different size population bases—with the four core neutrals ranging from 5–8 million and Yugoslavia closer to 25 million—but their different levels of manpower coming available each year affects final recruitment into the armed forces. On that score Finland has only 35,000 recruits available each year; Switzerland, Austria, and Sweden are in the 50,000–65,000 range; and Yugoslavia has nearly 200,000 of conscript age. Some might argue that such levels matter relatively little, since sustained, in-depth civil defense based on the 17–45 age pool may make the operational difference. Nevertheless, over the long run the additions (or lack thereof) do have an impact. As Vetschera points out in his chapter on Austria, serious difficulties have arisen for hitting the 300,000 reserves target level by virtue of too few new recruits receiving training.

The portrait we have is of these neutrals occupying the middle ground of many European country statistics. If they were larger, they might be obliged to carry their subregions with them in security terms. If they were smaller, they probably would not survive as neutrals. Such is the source of some of the questions about Austria today, being on the lower end, statistically, among the neutrals.

In terms of financial influence the limited budget resource of neutrals has a real impact on weapons acquisition. Jarvenpaa points out, in his paper on Finland, the limits on technological sophistication of domestic production. Thus tanks and interceptor aircraft have to be imported, creating delicate political issues. Only Sweden, among the neutrals, is immune to that problem at a higher threshold. And other neutrals are at an even lower threshold than Finland.

Philip Windsor points out the lack of a homogeneous neutrality in Europe. The existence of the five states under study here, plus Ireland, makes it inevitable that neutrality will be practiced in a variety of ways. It also means that some countries may become less neutral over time. For Ireland, as Windsor points out, the temptation to abandon neutrality comes not from military criteria, but rather from the economic reality of membership in the EEC. At what point does Ireland give up its posture of neutrality? And if Ireland already feels such pressure, what will happen to the other Europeans as 1992 approaches? Will those left outside the EEC fence suffer an economic penalty? Each of these temptations to abandon neutrality makes it that much more difficult to work as a bloc.

There is, too, a cultural difference in the willingness of neutrals to work as a bloc. Windsor notes the apparent instinct of Nordic neutrals to cooperate more readily than do those in south-central Europe. Without being spelled out, that degree of cooperation suggests the Nordics make a stronger contribution to what Lothar Ruehl calls "compensatory stability." The relationships between neutrals and alliance members are clearly more actively pursued on the northern front, even if still on an informal basis, due in large part to a kind of regional identity that does not exist so clearly elsewhere in Europe.

Neutrality has also had a major impact on the strategic designs of these countries. Clearly, neutrality implies no membership in defense alliances. It also requires a degree of self-reliance in strategic planning—something that members of alliances can afford to forgo. There is no "nuclear umbrella" for the European neutrals, and given the prohibitive costs of nuclear systems, all four have become antinuclear. And yet the prevalence of nuclear weapons all around them means that the neutrals cannot remain aloof from such issues. Their independent planning capabilities must be predicated on a level of resources available in the indigenous economy, such that if the neutral cannot produce its weaponry, it must at least be able to buy it from reliable sources. The escalating cost of weapons systems in the postwar period has thus been of great concern to neutrals. The changing technology of conventional warfare is creating financial requirements for the neutral states that could not be imagined a decade ago. At the same time the force levels in Europe remain extremely high, as they have been since the close of World War II. In this situation the neutral state's strategy must allow for its own ability to carry its defense burden.

At another level the neutrals must have a security design. Once a strategy has been adopted, where will it be pointed? No neutral state representative will discuss the identity of its enemies in public, aside from saying that they can defend their borders against all enemies. For some governments this calculation is more complex than for others. For instance, Finland has legal obligations that require its security design to be pointed in several directions. Others, such as Switzerland, have to worry about the counterstrike in any European war as much as it has to worry about the initial strike. But the theme of neutral defense, above all, is the sanctity of the national territory and defense of the borders.

Finally, there is a level of military behavior that occurs for neutral states that changes from time to time and cannot be acknowledged. Insofar as any neutral is able to define the likely enemy in its security design, it will engage in periodic, and generally not institutionalized, consultations to gain security from its neighbors. Even neutrals have to work out ground rules with neighbors in the security field, and these consultations can cover a wide range of contingencies. The longer that the broader geopolitical system remains static, the more comfortable the neutrals can be in their informal contacts with neighbors.

One cannot resist mentioning the peacetime influences that create alignments for the five European core neutrals, even as they maintain their neutrality. Their cultural identification and the intertwining of their economic fates with Western Europe have predisposed these countries to work increasingly with the West. The increasing openness of the Soviet economy under Gorbachev is meant to offset some of those attractions. The institutions of NATO may not tempt them, but it is clear that they identify, both for historical reasons and in order to build a future, primarily with the western part of the continent. As a result, wherever the legal construction or strategic logic may lead them in policy design, there are underlying forces at work that create intrinsic ties with the West.

The question of whether more neutral states are emerging is not directly addressed in this volume. Windsor raises the experience of the West with Hungary in 1956, and the disquiet caused in the East by that experience as well as the possibility that burgeoning neutrality could spill over into the East. But given the unrest emerging on the margins of Soviet influence during the current liberalization phase, it is not clear that it is in the Western interest to encourage discussion of more neutral states in Europe. Nevertheless, it is an issue with which we must engage, without knowing what course security issues will take in Eastern Europe.

There are many developments that are likely to cause the neutrals to broaden their dialogue and look jointly at issues and problems they all face, either as neutrals specifically or as members of the wider community of European states. This volume is intended to explore security aspects of those issues, so that both NATO members and neutrals can better understand the neutrals' contributions to European security.

PART ONE

THE NEUTRALS

CHAPTER ONE

SWEDEN

Lars B. Wallin

"Sweden's security policy, like that of other countries, aims to preserve the country's independence. The goal of our security policy should therefore be, in all situations and by means of our own choice, to ensure national freedom of action in order that within our own borders we may preserve and develop our society in political, economical, social, cultural and all other respects, according to our own values, and in conjunction with this to promote international détente and peaceful development." Thus were the goals of Swedish security policy defined by the Swedish Parliament in 1968, and so they were reconfirmed in 1972, 1977, and 1982, and once again when the Parliament voted on the latest five-year defense plan in 1987. In other words, national independence and freedom are the supreme goals, the policy of neutrality one of the tools by which Sweden hopes to assure them.

The Swedish policy of neutrality is a self-chosen policy; it is not laid down in the constitution nor internationally guaranteed. The present concept of Swedish neutrality has, in fact, gradually evolved over the last century and half; and although it was not always a consistent policy, its roots can be traced to the early nineteenth century.

THE EVOLUTION OF
SWEDEN'S POLICY OF NEUTRALITY

Séparés du reste de l'Europe, notre politique, comme notre intérêt, nous portera toujours à ne jamais nous immiscer

dans aucune discussion étrangère aux deux nations scandi-
naves; mais mon devoir et votre dignité serons toujours la
règle de notre conduite, et l'un et l'autre nous prescrivent
de ne jamais permettre qu'on intervienne dans nos affaires
intérieures. (Karl XIV Johan 1818)

[*Translated roughly:* Separated from the rest of Europe, our
politics, as well as our interest, leads us never to involve
ourselves in any quarrels which do not concern the two
Scandinavian nations; but my duty and your dignity will
always be the rule of our conduct, and both the one and the
other prescribes never to allow any interference in our internal affairs.]

When, in 1810, the French Marshal Jean Baptiste Bernadotte
was elected crown prince of Sweden, this was done partly in the
expectation that he would restore Finland, which had been lost to
Russia the year before, to Sweden. Bernadotte, who already as crown
prince rapidly seized control of Swedish foreign policy (he acceded
to the throne only in 1818), instead joined Russia in battle against
France and her allies, and in 1814, in the last war before a period
of peace which has now lasted for more than 170 years, he succeeded
to prize Norway away from Denmark. In 1834, when war was
expected between England and Russia, Bernadotte, now as King
Karl XIV Johan, in a secret note to these powers, declared his firm
intention to stay neutral. In a speech to the secret committee of the
Riksdag at the same time he expressed security policy views which
in many ways resemble present Swedish policy.[1]

While Karl Johan had conducted a policy that carefully balanced the interests of England and Russia and had emphasized
improved relations with the traditional archenemy, Russia, based
on "the absolute equality that is the foundation of all relations
between independent Souverains and which must never be measured
according to differences in their power," his son Oscar I and grandson
Karl XV, supported by activist liberal opinions, entertained ambitions of restoring Swedish influence in the north of Europe. In 1848,
for example, an army corps was deployed to Scania, the southernmost
Swedish province, and some thousand men were moved to the Danish

island of Fyen, in order to support Denmark in her war with Prussia over Schleswig and Holstein. Oscar I issued a declaration of neutrality on the eve of the Crimean War, but had the main theater of war been in the Baltic area and not in the Black Sea, he would have been prepared to join the Western powers against Russia. Dreams of reintegrating Finland were still entertained in some Swedish circles. In 1855 Oscar promised not to cede Swedish territory to Russia in exchange for English-French promises of support in case of a Russian attack (the November Treaty). When a new Dano-Prussian war approached, in 1863, Karl XV promised the Danish king a defense union, but his cabinet objected and the plan had to be canceled.

With Prussia's victory over France in 1871, those Swedes who were still dreaming of the resuscitation of Sweden as a great European power lost their last hopes. The Baltic region was now dominated by a Russia suspicious of Swedish intentions and a Germany whose feelings towards Sweden were less than warm. It became the task of Karl's brother, Oscar II, to improve relations with these powerful neighbors.

In the early years of the present century there were widespread fears in Sweden of a Russian attack through northern Scandinavia, either in isolation or in connection with a larger European war, for the purpose of accessing open waters. In 1910, with the permission of the government, a representative of the Swedish general staff conducted discussions in Berlin concerning the possibility of German assistance in case of such an attack. While Germany was quite prepared to exploit pro-German sentiments in Sweden to make Russia deploy larger forces to northern Europe, thereby providing Germany with more freedom of action on the Continent, she had no interest in giving Sweden any guarantees. However, these talks, as well as the general anti-Russian sentiment in Sweden and some rather ambiguous statements by the Swedish foreign minister when war broke out in 1914, may have contributed to Russian suspicions as to the reliability of Swedish neutrality. One famous incident from the first days of World War I may have been caused by such doubts and it is often used to demonstrate the importance of credibility to a policy of neutrality. On 9 August 1914 the Russian Baltic Fleet left its base in Helsinki and steamed towards the Swedish island of

Gotland, with the intention of sinking parts of the Swedish navy near Fårösund. It is not known whether the Russian command doubted that Sweden was about to join the war on Germany's side, or if it was uncertain of the accuracy of reports that Swedish and German warships were joining up in the Baltic for attacks on Russian naval units and harbors, but fortunately von Essen and his Baltic fleet only got halfway before being countermanded by St. Petersburg.

As it turned out, the First World War caused no military problems for Swedish neutrality. Instead, it became to a large extent a question of defending a neutral's rights of unimpeded trade. What problems there were of a military nature had to do with Finland. In 1917, after the Russian Revolution, Finland had gained its independence from Russia, and in January 1918 civil war broke out in the newly formed republic. Finnish requests for help caused a number of difficulties for the Swedish government. Although neutrality was not properly at issue—Finland was not at war with any external power—there was still a definite risk of becoming entangled in the ongoing war—or so it was perceived in Stockholm. Furthermore, Swedish political opinion of the warring sides in Finland was very much divided. The Finnish requests were turned down, although permission was given for the transit of some war materials, German transports were escorted to harbors in the Gulf of Bothnia, and some Swedish officers were allowed to volunteer on the government side.

Another problem was caused by the eternal Åland question. Åland is an archipelago joining Sweden and Finland across the northern end of the Baltic. Belonging to Finland, these islands were for a long time a political and strategic sore spot for Sweden—politically because their population is entirely Swedish-speaking, and strategically because who controls Åland controls the access to the Gulf of Bothnia and, consequently, all shipping along the northernmost two-thirds of Sweden's east coast. In the hands of a hostile power they would also be regarded as a serious threat against the Swedish capital.

In the autumn of 1917 the islands' inhabitants made known their wish to detach themselves from Finland and join Sweden. Germany offered Sweden help to acquire the islands. Although the

Swedish government did not take a definite stand at the time, following reports of Russian persecution of the population a detachment of six hundred men was landed on Åland to disarm the Russian forces still present there, and to supervise their repatriation. This intervention was not appreciated either by Finland or by Germany, and the Swedish unit, the first one to be dispatched outside Sweden since 1849, left Åland in May 1918. Following a ruling by the League of Nations, Åland was assigned to Finland, but home rule was guaranteed. Following an international treaty of 1921, Åland was demilitarized and neutralized. This settled the matter for a while, but only until a new war was in preparation.

The period preceding the next war was a time of optimism, symbolized by the League of Nations. Although it was questionable whether membership was consistent with neutrality, in particular because of the clause on collective military sanctions, Sweden opted for membership without stating any formal reservations (in contrast to Switzerland, which was granted exemption from participation in collective military actions). In fact, for a while "neutrality" vanished from the Swedish foreign policy vocabulary. Instead, references to international law and solidarity between nations as safeguards for the rights of small states became salient in official pronouncements.

The League of Nations period was also a period of great hopes for disarmament, and Sweden was quite active in this work, with her first Social Democratic prime minister, Hjalmar Branting, and his party friend Rickard Sandler playing prominent roles. Soon, however, illusions were dispelled and preparations for a new war got under way.

In 1938 all four Nordic countries issued declarations of neutrality and they also explored, in a preliminary fashion, the possibilities of cooperation in the safeguarding of Nordic neutrality. However, their perceptions of potential threats were too different for these discussions to lead to any real results. A limited agreement of cooperation was concluded between Sweden and Finland concerning a joint defense of Åland—the Stockholm plan of July 1938. To some extent this was motivated by the Swedish interest in not allowing a hostile power on these islands, but it was mainly perceived as a way of anchoring Finland to a policy of neutrality. Detailed plans for the defense of Åland were ready in April 1939, but when the question of implementation was raised by the approaching Finnish Winter War it all came to naught.

When the Winter War broke out on 30 November 1939 Sweden took a nonbelligerent position. Finnish hopes for Swedish intervention did not come true: Swedish support was limited to deliveries (albeit substantial) of war materials, including aircraft,[2] a volunteer corps of approximately eight thousand men, loans, and humanitarian aid. On 12 March 1940, one day before the end of the war, Norway and Sweden declared themselves willing to discuss a defense union with Finland. Negative Soviet reactions quickly aborted this initiative, however, and the German attack on Norway and Denmark on April 9 sealed its fate. In October, on the initiative of the Finnish government, discussions of Swedish-Finnish cooperation were initiated, even a formal union with its foreign policy conducted by Stockholm was among the alternatives envisaged. Again, Soviet and German hostility put an end to the discussions.

The real test of Swedish policy was, of course, the Second World War. From very early on, and until the last days of the war, Sweden was surrounded by German troops. I will save a discussion of some incidents relating to Swedish neutrality in this war for later.

In 1946 Sweden joined the United Nations, again despite questions as to the compatibility of neutrality with membership in a collective security organization. Contrary to the situation in the old League of Nations, the UN Charter makes participation in collective military interventions mandatory, when decided upon by the Security Council. However, Sweden's main concern—to stay out of conflicts between the major powers—was felt to be satisfied by the veto powers of the permanent representatives of the Security Council.

In 1948 Denmark, Norway, and Sweden discussed a Nordic defense union. While Denmark showed a strong interest, Norway did not feel that a neutral block of the kind envisaged by Sweden would correspond to its security interests, and by the end of 1949 the present Nordic pattern had been established, with Denmark and Norway members of NATO but with well-known restrictions on foreign bases and nuclear weapons, a nonallied Sweden, and a likewise nonallied Finland. The latter country had, after the trauma of war, managed to achieve a situation of mutual trust and friendly relations with the Soviet Union, symbolized by the Treaty of Friendship and Cooperation of 1948.

THE CHARACTER OF SWEDEN'S POLICY OF NEUTRALITY

Although often used as a convenient shorthand, the term neutral is, strictly speaking, appropriate only for Austria and Switzerland, which are permanently neutral states. Finland and Sweden, on the other hand, both conduct *policies of neutrality* in peacetime aiming at neutrality in the event of war. Obviously, membership in a military-political alliance would not make such policies credible or viable, thus the commonly used description of the Swedish policy as one of "non-participation in alliances aiming at neutrality in the event of war."[3] This is not a mere quibble about words; it is a way of expressing the intention not to allow any outside power to prescribe what is consistent or not with Sweden's policy of neutrality. Statements such as "we are the sole arbiters of our policy of neutrality" are sometimes ridiculed, and were they indicative of a belief that Sweden can do whatever it pleases and still maintain credibility of its intentions to stay neutral in the event of a war, they would be rightly ridiculed. On the contrary, Sweden is acutely aware of the crucial importance of conducting a credible policy of neutrality, but also of the necessity to safeguard freedom of action—to not allow any foreign power to use neutrality arguments as a means of influencing Swedish policy.

The purpose of the Swedish policy of neutrality is to make it possible for Sweden to stay out of a war: nothing more, nothing less. This implies that the policy be conducted so as to convince everyone that Sweden firmly intends to avoid becoming involved in a war between other countries. It must give rise neither to fears nor to expectations on the part of either of the major powers. It does not imply that Sweden is politically equidistant from the powers, nor that its opinion is neutral. Politically, socially, and economically Sweden is firmly entrenched among the Western democracies. This does not rule out occasional criticism of both East and West in cases of perceived infringements of international law. The Soviet interventions in Czechoslovakia and Afghanistan, and the American in Vietnam and Central America are cases in point. Sweden takes a very strong interest in the defense and the strengthening of international law, not least as regards noninterference in the affairs of and the right of self-determination of all states.

SWEDISH SECURITY POLICY: THE SECURITY ONION

If the policy of neutrality is a means to an end—the safeguarding of freedom and national independence against the most serious threat of all, war—it is not the only means. Looking at Swedish security policy as a whole, one can discern several levels or layers, each of which employ different tools and have goals that lie in different time frames. The goal on the first level is to improve Swedish security by the reduction of the risk of war overall. Measures to promote a just global society are part of policy on this level. Although I agree with the late Swedish economist and social scientist Gunnar Myrdal that the Swedish development aid target of 1 percent of the GNP is founded on a strong sense of solidarity with the Third World by the Swedish people, foreign aid is also regarded as part of Sweden's security policy. It is hoped that, by relieving social and economic tensions due to unequal development, foreign aid can decrease the risk of conflict in the Third World with possible superpower involvement and a consequent risk of conflict spreading to Europe. The active Swedish interest in a New Economic Order as well as in the activities of the United Nations and its agencies can also be seen as expressions of this belief.

As a more direct way of preventing the spreading of crises, Sweden has a long tradition of participating in different UN and other peacekeeping activities: mediating efforts (Count Bernadotte in the Middle East in 1948 and Ambassador Jarring in the 1970s; the late prime minister Olof Palme in the Iran-Iraq War in the 1980s); supervisory commissions (UNTSO [Middle East, 1948–present], UNMOGIP [India-Pakistan, 1949–present], NNSC [Korea,

1953–present], UNOGIL [Lebanon, 1958], UNYOM [Yemen, 1963–64], New Guinea [1962], UNIPOM [Kashmir, 1965–66], UNMOGAP [Afghanistan, 1988–present], UNIIMOG [Iraq-Iran, 1988–present]); UN peacekeeping forces (battalion size units in UNEF [Gaza, 1956–67], UNOC [the Congo, 1960–64], UNICYP [Cyprus, 1964–present],[4] and in Sinai [1973–80]). Since 1980 Sweden has contributed a medical company to UNIFIL in Lebanon. A special emergency force for UN peacekeeping operations was organized in 1964, concurrent with similar forces in Denmark, Finland, and Norway. Consisting of two infantry battalions and special units of maximum battalion strength, and equipped with light arms and vehicles, this force, if requested by the UN and after a decision by the Swedish government, can be dispatched on short notice to wherever need may be.[5]

The next level of security policy has to do with East-West relations. Efforts to promote détente rank high in Swedish security policy. Swedish disarmament policy is probably to a large extent motivated by a concern for détente, although arms control is also seen as reducing the threat of war and its consequences. The proposal for a battlefield nuclear weapons-free corridor in Europe was probably perceived as contributing toward all of these goals. For a small nonallied European nation, disarmament could also be conceived as a strategy for relieving its defense burden or for improving its military position relative to the power blocs. However, Swedish efforts have hitherto been directed toward obtaining a chemical weapons ban, a nuclear test ban, and reductions in nuclear arms rather than toward obtaining reductions in conventional forces. Sweden has, for example, been an active member of the Conference on Disarmament in Geneva since 1962.

In Swedish threat perceptions, isolated attacks have a very small place, if any. A war involving Sweden is generally believed to be part of a larger European conflict. Measures aimed at reducing tensions and the risk of war in Europe therefore rank high on the Swedish foreign policy agenda. A concrete expression of this was the Swedish offer to host what became officially known as the Stockholm Conference on Confidence- and Security-Building Measures and Disarmament in Europe.

While Sweden's means of influencing the military and political situation in Europe as a whole are limited to those of foreign policy and diplomacy, the credibility of her policy of neutrality and the strength of her total defense are believed to be vital factors in the stability of the Nordic area. The goal of maintaining northern Europe as an area of low tension and low superpower military presence has taken a prominent place in Sweden's political and defense planning agendas.

On the penultimate level Sweden adheres to a strategy of conventional dissuasion. Based on the assumption that her territory is seen mainly as a transit or base area for military operations directed at targets in the vicinity, Sweden hopes that her defense forces will dissuade a prospective aggressor by convincing him that the price for trying to use her territory would be too high. The ultimate level of Swedish security policy is, of course, direct defense, should the conventional deterrence strategy prove unsuccessful.

The destructiveness of a modern war, and the disastrous consequences, not only for the warring parties but also for the bystanders, of any European war which involved NATO and the Warsaw Pact, even if limited to conventional weapons, has made the avoidance of war a prime goal of Swedish security policy. Peacetime diplomacy and regional crisis management therefore often tend to get more emphasis in the political discourse than measures directed at wartime problems. In a 1985 article Heinz Vetschera suggests that Finland regards foreign policy as her predominant security policy tool, while Sweden and Austria assign equal weight to foreign policy and defense, and Switzerland emphasizes defense.[6] Some time ago, the relation between defense and foreign policy was, in fact, much discussed in Sweden. This debate, I believe, was not so much about security policy as about domestic and election politics, and it died more or less after the Defence Committee issued its security policy report in the spring of 1985. Seen in the broader perspective of the overriding importance of avoiding a war involving the superpowers or their allies in Europe, the supremacy of foreign policy over defense in Swedish security policy could very well be argued. In the more limited context of neutrality policy it is very much a debate about the pope's beard. In this context foreign policy and defense are clearly symbiotic.

Although Sweden sets great store by international law, she is under no illusion that this is an area which is characterized by a very high degree of either law-abidingness or enforcement. This is not to say that, over time, views on acceptable behavior do not change, become assimilated, and influence even wartime behavior, or that international law cannot limit options in war by restricting peacetime preparations which would be inconsistent with international norms, or that reasons of self-interest cannot be invoked for abiding with the laws of war even in situations when breaking them would appear expedient. In World War II, for example, decisions against the use of chemical weapons in the Pacific campaign may have been influenced by the fear that this would provide Germany with an excuse for employing chemicals against vulnerable invasion forces in Europe.[7]

While in the nineteenth century military forces and neutrality seem to have been regarded as contradictory, the opposite is true today.[8] The Fifth Hague Convention obliges a neutral to maintain control of its territory and to defend against infringements of its neutrality. However, while the neutral is obliged to use the means at his disposal to the best of its ability, the exact strength of his means is not prescribed. In the final analysis neutrality, and a policy of neutrality, although based on the rules of international law, is, and has to be, approached in a pragmatic manner. In this connection I quote Ambassador Anders Thunborg, who, speaking as Swedish minister of defense to the International Seminar for Diplomats in Klessheim, Austria, in 1984, observed that

> needless to say, a belligerent state will care little for neutrality, if the advantages of an attack are considered to outweigh the disadvantages resulting from violating the rights and interests of a neutral state. A neutral state lacking an adequate defence will hardly be able to stand up to the crucial test of credibility. . . . Nobody is or will ever be in a position to guarantee that Sweden simply by pursuing a policy of neutrality will be spared the dire consequences of war. This is exactly why we maintain and will continue to maintain defence forces sufficiently strong to make any

potential aggressor think at least twice. . . . [But] no
amount of military power will suffice to guard the peace, if
the foreign policy pursued by a neutral state does not inspire
and sustain the confidence of the surrounding world in its
determination and ability to stay free of commitments that
might jeopardize its neutrality in time of war. There must
be no doubt about its determination to guard at all times its
independence and freedom of action. Above all there must
be no expectations nor any fears that the policy of neutrality
will be changed due to external pressures.

The warring parties must be reasonably sure, not only that they
will not be able to use the neutral's territory without having to pay a
substantial price in blood and time, but also that their adversaries will
experience the same fate should they attempt it. Thus, a sufficiently
strong defense is necessary for the credibility of neutrality; but unless
combined with a credible policy of neutrality, strong military forces
conceivably could be perceived as a potential threat by one or the other
party. Clearly, a country aspiring to neutrality in war must avoid ties
that would significantly limit its freedom of action in war, such as
international cooperation implying mandatory political consultations
or joint decision making on foreign policy. Although the state of neutrality is terminated if a neutral state is attacked, and although it is
then likely to join forces with the other side, peacetime preparations for
defense cooperation, including staff talks such as took place between
Switzerland, on the one hand, and Germany and France, on the other,
before World War I, and between Switzerland and France before
World War II are, in the Swedish view, definitely ruled out.[9] This has
nothing to do with any peacetime legal restrictions on a policy of
neutrality, only with its credibility.

Two questions remain: Is neutrality a real possibility in a future
war? and, Is the Swedish policy of neutrality credible? Before turning
to the first question I discuss some aspects of the second one.

DIMENSIONS OF CREDIBILITY

Although Sweden and Switzerland were the only neutrals on the
continent of Europe to emerge from the Second World War with

their neutrality intact, they had certainly lost their innocence. Both countries found their freedom of action severely restricted and both were forced to some measure of accommodation when the balance of power was thoroughly upset in their respective neighborhoods. For Sweden the war had been traumatic in several respects, not least for its national pride, and several lessons were learned. One was the importance of a strong defense as a support for the conduct of an independent national policy.

Following the defense decision of 1925, the military establishment was drastically reduced. The defense budget remained at essentially the same level, though, because at the same time the armed forces were modernized, and in 1926 an independent air force was established. In the mid-thirties a slow buildup began, quickly followed by a drastic acceleration in defense spending as the war drew closer. In 1941 11.8 percent of the Swedish GNP was spent on defense. Although at a reduced rate, defense expenditure continued to rise, and in the 1950s Sweden could boast one of the largest air forces in the world.

Today things are different. Still, however, with roughly 3 percent of its GNP devoted to its total defense efforts, Sweden leads the industrialized West European neutrals in this measure, and compared with the West European NATO countries she falls somewhere in the middle. In real terms defense spending peaked around 1980. A slow decline followed, but this has probably been arrested and reversed to a moderate increase with the five-year defense budget that was adopted in 1987.[10]

Except for a rather small cadre of professional officers (this includes NCOs in present Swedish terminology), the Swedish army is entirely based on conscript forces. Having no ready units (with the exception of training units close to the end of their basic training and units recalled for some weeks of refresher training), the army should be able to mobilize 700,000 men in seventy-two hours. This includes a field army of 300,000, consisting of five armored/mechanized brigades, eighteen infantry, and five Norrland brigades,[11] and some seventy independent infantry, ranger, armor, artillery, and air defense battalions; local defense forces of a similar size with some ninety battalions, and about four hundred independent companies; and a home guard of 120,000 men.

Compared to the central European armies, the Swedish army's heavy weapons inventory is not very impressive, nor is it very large in relation to the size of the country. When making such comparisons one must, however, take account both of the size of the threat and, in particular, the very different circumstances of terrain, geography, and climate which characterize the north of Europe, especially northern Sweden. The brigades are gradually being modernized. Financial restrictions will, however, limit the number of fully modernized ones. A new 155-mm howitzer is being introduced, air defenses are being strengthened by the progressive introduction of the laser-guided Rb-70, the battle tanks are due for refurbishing, and a new generation of armored fighting vehicles is being developed.

Relying on a mobilization army is, of course, somewhat problematic, particularly in a country with long and vulnerable lines of communication and a small, very unevenly distributed population. Still, a very dispersed pattern of mobilization depots, typically storing all the equipment of a company, including uniforms and other personal kits, as well as light and heavy weapons and vehicles, does give the system reasonable resilience. Mobilization is protected by very easily mobilized home guard units, and particularly by an air force which is essentially at full readiness.

The Swedish air force today numbers eleven air defense squadrons, most of which are equipped with the very capable JA37 Viggen interceptor (some 220 aircraft in all), five medium attack squadrons with JA37 and five light attack squadrons with the SK60 trainer (some 150 aircraft), six recce squadrons with SR37 (some 55 aircraft), and auxiliary units (transport, liaison, rescue helicopters). In terms of numbers of aircraft, the air force has suffered heavily from the drastic cost increases for modern fighter aircraft; from the point of view of quality, though, it is very much in the forefront among Europe's air forces. An already very dispersed system of war bases and reserve airfields is being further expanded and will be completed during the 1980s. A new generation Swedish-built multirole fighter, JAS39 Gripen, will be introduced in the early 1990s.

The Swedish navy has been forced to reduce its ship inventory.

The last two of around twenty destroyer/frigates in existence at the end of the 1960s were recently decommissioned, and the number of submarines has fallen from some twenty to twelve in the same period.[12] Better weapons and new classes of submarines have, however, maintained the effectiveness of the Swedish submarine fleet despite reduced numbers.

The elimination of the destroyers and frigates was partly the result of a financial squeeze. It also resulted from a realization that large surface combatants are too vulnerable in restricted seas like the Baltic and that smaller craft with significantly improved firepower would be more useful. Instead emphasis was put on exploiting the natural defensive advantage of the extensive archipelagos which cover large stretches of the Swedish coast, using mines and fixed and mobile coastal artillery for coastal defense, fast torpedo and missile craft, operating from protected positions among the islands, for rapid attacks on the open sea, and relying on submarines and antiship missile equipped attack aircraft as a forward screen.

The elimination of the destroyers and frigates also meant that most of the anti-submarine warfare capability was lost. This was not seen as particularly ominous, as no need for extensive convoy missions was perceived for the future. The recent submarine incursions have, however, caused the buildup of a shallow-water ASW capability of a kind that has never before existed in the Swedish navy.

Having the military wherewithal is one thing (and I do believe it is reasonable to say that the Swedish defense forces would be able to extort a very high price from any attacker), but being both able and having the determination to use it is undoubtedly something else. This relates to another lesson of World War I (and of World War II too): the importance of an economic defense—the ability to sustain the economy, to have access to world trade, and to be able to clothe and feed the population, without having to yield to foreign pressure. Also important to a credible defense is the ability to physically protect the population. Increased efforts are being directed at finding and implementing measures to decrease Swedish dependence on foreign imports in wartime and to improve civilian protection. Lately there have been considerable efforts to identify ways of mitigating the special vulnerabilities of a modern society that is dependent on large technical systems for the satisfaction of most basic needs. This is all part of the Swedish Total Defence concept. One indication of the emphasis on the adequate preparation of society for crisis and war is the reorganization, on 1 July 1986, of the civilian branch of the total defence, when the National Board of Economic Defence became the National Board of Civil Emergency Preparedness, with a director general who more or less will assume the role of "Supreme Commander of the Civil Total Defence." Another indication is the rapid increase in the volume of research at the Swedish Defence Research Establishment on the interaction between military defense and the civilian parts of society in crisis and war.

But, of course, all this is organization and planning. Ultimately any defense of a nation must rest on the will and determination of its population to resist aggression. Short of a very serious crisis there is no way to measure this except by opinion polls. Polls related to defense questions have been carried out on behalf of the National Psychological Defence Planning Committee since the early 1950s, at first at irregular intervals, at least once a year from 1965 on. While suffering from the usual weaknesses of opinion polls when it comes to elucidating the population's "true" beliefs and opinions, and, in particular, as a guide to its actual behavior when the chips really are down, this constitutes quite an extensive data base, which at the very least can be used as an indication of how views on defense issues in Sweden have varied over time.

While positive to disarmament, if global, the Swede does not favor unilateral disarmament (fig. 1.1).[13] Swedes are 90 percent unanimous that a defense force is necessary. As shown in figure 1.2, Norwegian opinion is closely similar. And according to an international comparison from 1983, the Swede appears more prepared than most to take up arms against an aggressor (fig. 1.3).

"Assume that Sweden is being attacked. Is it your opinion that we should resist by force of arms even if the odds are uncertain?" is one of the questions which have been posed in the polls since 1952. Figure 1.4 shows that the result of the international study was not a spurious one but characteristic of Swedish postwar opinion. While the age group of 25–29 years is somewhat less positive (68 percent

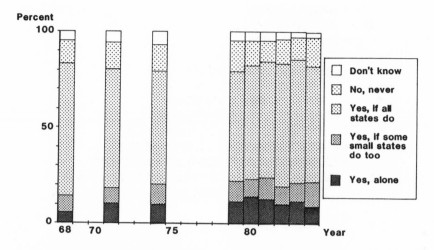

Figure 1.1 Should Sweden Disarm?

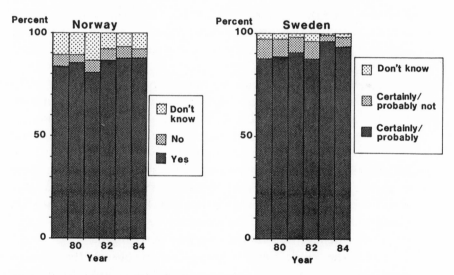

Figure 1.2 Is a Defense Necessary?

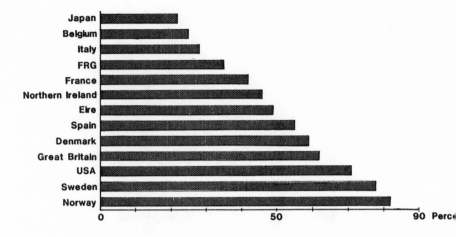

Figure 1.3 Comparative Degrees of Defense Resolve

in 1984), and the oldest group (65–70 years) the most positive (89 percent in 1984), it is interesting to note that the youngest group (18–24 years), which includes males who are in the midst of, or have recently completed, their basic military training, is close to the average.

Figure 1.5 reveals an increasing perception among Swedes of a need to strengthen the defense. Polls taken at a higher frequency for a period after the well-publicized foreign submarine incursions in the 1980s showed that the opinion was very sensitive to the ups and downs of submarine activities along the Swedish coasts; but from the diagrams presented here it is evident that these fluctuations were superposed on a clear long-term trend (one should note that roughly two-thirds of the population have consistently believed that Swedish chances of a successful defense are rather or very small, while 25–30 percent believe that they are very or fairly high).

That the Swede seems to be willing to consider the financial consequences of improving the defense can be concluded from figure 1.6; never, in the past thirty years, have so many been in favor of increased defense spending and so few argued for decreased spending, as in the past few years.

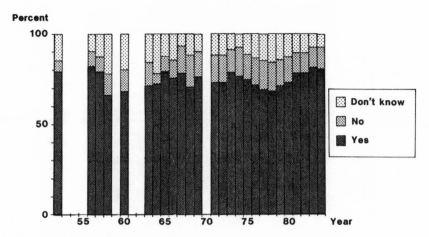

Figure 1.4 Attitudes on Armed Resistance

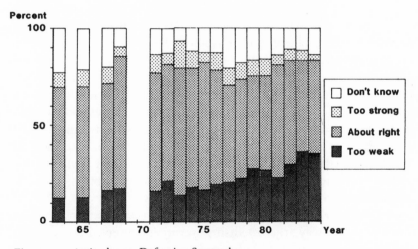

Figure 1.5 Attitudes on Defensive Strength

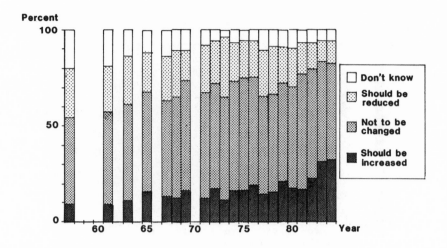

Figure 1.6 Attitudes on Expenditures for Defense

Again, this still does not prove the credibility of Swedish neutrality; there can be no such proof. Those who are inclined to do so might turn to history and reflect over its possible lessons, particularly what lessons Sweden might have drawn. But this is a dangerous exercise: relevant facts may not be known, there is no guarantee that circumstances are the same, ambiguities abound, and there is much room for interpretation and for selective memory.

Without getting too involved in the complexities of the Second World War situation in Scandinavia, I bring up some events which I believe consciously or subconsciously influence perceptions of Swedish behavior in possible future contingencies.

In the early morning of 9 April 1940, as German troops were landing in Norway, the Swedish government received a note from the German government containing assurances of no hostile intent, while demanding strict Swedish neutrality and the abstention from all military countermeasures. Mobilization, which had been slowly initiated some days earlier, continued, however, and by the end of the month some 330,000 men, albeit not very well equipped, stood at arms in Sweden.

It is common knowledge that, following German demands,

Sweden consented, on 19 July 1940, to certain German transports on Swedish railroads. This transit traffic was to be limited to German soldiers on leave, traveling without their weapons, and was to result in no change in the number of forces in Norway. Outside this agreement, consent was also given to transports of very limited numbers of troops between Trondheim and Narvik over Swedish railways, the so-called horseshoe traffic.

What is less well known, however, is what had—or had not—gone on before this final concession. Since the second week of the war in Norway, the Germans had repeatedly requested permission to transit military supplies to their forces in Norway, in particular to those around Narvik. While some transports of food and sanitary equipment as well as of a number of medical orderlies were approved, Sweden firmly rejected demands for the transit of war materials and also exposed German attempts at smuggling weapons in Red Cross transports. It should be remembered that Sweden (and Finland) by then was entirely surrounded by German forces or by nations aligned with Germany, and that it depended on Germany both for weapons and for much-needed coal. Admittedly, Sweden had one trump card too, its iron ore (which was, in fact, what the fighting between British and German troops around Narvik was very much about). Some refusals were made despite a real fear of an imminent German attack and accompanied by military alerts. A number of German aircraft were shot down. On 28 May Allied forces captured Narvik, which made the situation still more threatening; Hitler was not prepared to accept defeat in this area. Britain had, however, already made the decision to withdraw from northern Norway. On 8 June hostilities ceased in Norway and there seemed to be no pressing need for German transits through Sweden. But on 16 June new requests were made, in a demanding tone, and on 18 June Sweden gave notice of its acceptance. By then the strategic situation had changed drastically, not only in the north. Sweden could no longer claim that allowing transits would equal stabbing a neighbor and friend in the back, an argument which had had some influence with Göring on an earlier occasion. Italy entered the war on 10 June, France agreed to a cease-fire on 16 June. In case of war with Germany Sweden would be in a militarily hopeless situation, with no prospects of help. In addition, a telegram from the Swedish embassy in London reported indications of a British desire to conclude peace with Germany.[14]

Another well-known incident was the transit, on German and Finnish insistence, of a German division, Division Engelbrecht, in June 1941, from Norway to Finland. Following a very hard debate in government and Riksdag, permission was granted. Interestingly enough, opposition to the concession was strong, particularly among the Social Democratic members of the government coalition and the Riksdag. Among those most heavily opposed were two men, Rickard Sandler,[15] who had been known as a strong proponent of disarmament in the League of Nations, and östen Undén, who, as long-term postwar minister for foreign affairs, to a large extent designed the modern Swedish policy of neutrality. The Swedish government publicly emphasized that the transit had been definitely declared by both parties to be a onetime measure, and a new German request, on 31 July, was turned down. Although a clear breach of neutrality, the concession obviously was not regarded to be of much military significance, and it elicited only comparatively weak protests from the Allies. There was no formal British protest, although definite warnings were given to the Swedish foreign minister lest the concession lead to new ones of increasing scope and frequency. The American government, to cite Carlgren, let it be known that it considered Sweden's foreign policy to be based on a realistic assessment of the current situation from a strictly Swedish point of view, and noted in particular the declaration that Sweden intended to resist by force of arms any military threat which endangered the country's sovereignty, integrity, or independence.[16] Even the Soviet Union, although delivering a formal protest, did so in a fairly low key and accompanied it by assurances that no deterioration in Soviet-Swedish relations was implied.

I end these notes with a quotation from a speech by Dr. Undén in the Riksdag, 16 August 1940.

> When a threateningly protective great power advances its really dangerous demands on a smaller state, the small state is sometimes receptive to the suggestion that even

very large concessions are better than war. But should a consequence of the concessions be that the small country voluntarily and without resistance gives its powerful protector a foothold within its borders, then it has accepted the policy of the Trojan horse in its coarsest form. What is going to happen next is that the small country will be robbed of even its ability to fight for its freedom. Defense cannot get off the ground, the military machine gets out of gear, the population becomes paralyzed. Of course, the result may be the same if the path of resistance is chosen from the beginning. But then the Nation has at least saved its soul. And future generations have been entrusted an obliging heritage.

IS NEUTRALITY POSSIBLE?

There are at least two ways of answering the question whether neutrality is possible. One has to do with the morality of neutrality, the other with its future practicality.

In World War II all warring sides used moral arguments against the stands of the neutral states, and when the establishment of the United Nations was discussed in San Francisco after the war it was even suggested that neutrality be outlawed in the UN.[17] Views have changed since then, although criticism of neutrality is far from uncommon, even today.

A standard Swedish response, when the moral question is raised, is that the responsibility of a government is first and foremost toward the security of its own people. One might add that few, if any, states are so altruistic that they invite war on their own territory for the sake of a moral argument. (Not to be willing, unprovoked, to invite war on one's own territory is, of course, not the same as not being determined to resist by force of arms any attempt by a foreign power to use one's land, sea, or airspace in operations against a neighbor.)

The question of the morality of neutrality should perhaps be put in a wider context than that of war. Peace is, after all, a more likely occurrence than war. In this connection it might be appropriate to invoke some words by Professor Hans J. Morgenthau, who in his *Politics Among Nations* concludes his exposition by enumerating some fundamental rules of diplomacy, which in his opinion ought to guide the conduct of nations if the national interest is to be promoted and peace preserved. Foremost among these rules is the proposition that diplomacy should be divested of the crusading spirit, as the invocation of abstract moral principles in foreign policy only serves to magnify conflict at the same time as it erodes the basis for necessary compromise. To help find this basis for necessary compromise is one of the valuable roles which neutral nations can play in international politics.

Turning to the practicality of neutrality, one has to recognize several factors. One factor is the character of possible future wars and changes in the strategic environment in general; a second is the ability of a neutral nation to maintain its defensive capabilities; and a third relates to the political/economic environment.

There is no doubt that a neutral country would fall victim to at least the collateral effects of a nuclear war in Europe; nor is there any doubt that it would suffer considerably by a large-scale conventional war. But, apart from the potential magnitude and character of the effects, this is not an entirely new situation. Furthermore, to be a participant in the war would hardly improve a neutral country's lot.

Technology has provided improved means for offensive operations in the Nordic environment. Increased range for combat aircraft reduces the protection afforded by distance; helicopters and fast naval or surface effect vessels increase the prospects of short warning attacks and to some extent reduce the advantages provided by difficult terrain and sea borders. As already noted, improvements in weapons technology and in mobility mean that Sweden can no longer count on being isolated from a war in the rest of Europe.

Whether the exploitation of this new technology also requires the involvement of Sweden or its territory is questionable. In fact, the opposite could be argued as well. It is, of course, imperative that these developments in technology and options be heeded when planning Sweden's defense strategy. As long as this is done, however, I cannot see that the situation has become qualitatively different.

After all, Sweden did not build up a comparatively strong defense just for its own sake. It was done because Sweden realized that foreign powers might consider it to be to their strategic advantage to gain access to Swedish territory, and that many powers have the wherewithal to attempt this.

But it is clear that the technical and strategic environment does change, that this influences Sweden's defense capability, and that it must be met by adjustments in technology, tactics, defense structure, and strategy.[18]

Such adjustments are partly a matter of the application of brainpower and planning, but they are also, to a very large extent, a matter of money. New technology is expensive—modern fighter aircraft and modern armor particularly so. Like other Western countries, Sweden was affected by the increased costs of energy in the 1970s; this was compounded by the effects of too liberal salary increases and a built-in explosion in social welfare costs. Savings in the state budget weakened defense as well as other public sectors. Savings were sought by reduction of the peacetime establishment: "rationalization" of the central staffs and reductions in the number of training units, cancellation of some refresher training periods, and slower modernization of the army field brigades. However, while the number of air defense squadrons was reduced, modernization of the air force continued apace, and the development of a new fighter for the 1990s got under way. Although the defense budget is once again increasing in absolute terms, economy will certainly continue to be a problem for this small nonallied country, with its large territory, which has to contend with the far superior technological resources of the superpowers.

Another aspect of the attempt at economy is the defense industry itself. Sweden has for a long time had a very well developed defense industry, fulfilling some 70 percent of the country's defense material needs. Although domestic production is often more expensive than imports, it has been regarded as a boon in several ways. It has allowed the development of equipment that is particularly adapted to Swedish climate, terrain, and conscript forces. It has diminished the security and neutrality policy risks of being dependent on foreign sources; and it has guaranteed Sweden's ability to maintain and service its own vital weapon systems in times of war. However, the increased costs of modern weapons, coupled with their dramatically improved capabilities, tend to augment the costs of domestic production out of proportion, due to the consequent reductions of the production series.

Extending the production series by foreign sales is possible only to a limited extent because of a rather strict export policy. The future of the Swedish military aircraft industry is thus very much in doubt. This is an area where competition is ruthless and where the decision to sell is not entirely in Swedish hands. A case in point is the attempted export of the Viggen fighter to India some years ago, which was vetoed by the Americans. (The Viggen engine is a modified American engine produced by Volvo Flygmotor on American license.)

Keeping up with technology is a third problem, although so far Swedish industry has managed quite well in this respect. Access to foreign technology is imperative, though, and the increasing restrictiveness, for commercial and security reasons, which characterizes the field of technology transfer could pose more and more serious difficulties. Continued militarization of technological research and development may put obstacles in the way of necessary cooperative efforts between Swedish and foreign industry.

The technology transfer issue is a problem also on the political or political/economic level. In order to gain access to modern technology, neutral and nonaligned countries have been forced to acquiesce to some measures which are not necessarily congruent with their security policy.

The increasing economic/industrial integration in Western Europe requires vigilance on the part of the neutrals, who must avoid becoming so dependent that the possibility of conducting an independent policy of neutrality becomes illusory. At the same time it is true that a strong economy makes the people both willing to support a strong defense and able materially to sustain it. It is also commonly accepted that in the future participation in European economic integration efforts is a prerequisite for a strong economy. So far Sweden has been fortunate to find modes of association with the European Community while avoiding becoming entangled in activities aimed at the formulation of common foreign policies. Participation in such aspects of European integration is considered wholly incompatible with the freedom of action necessary for a credible policy of neutrality.

It is imperative, from the point of view of the neutral and nonaligned nations of Europe, that West European integration continue to evolve in such a way that these nations can participate in its economic aspects, with full privileges and obligations consistent with a policy of neutrality, while being allowed to stand aside from foreign policy cooperation.

WHAT CAN BE DONE?

Nuclear war, even if limited to the theater level, would have serious consequences for everybody (belligerent or not) close to the theater of war. Furthermore, a limited nuclear war entails a high risk for escalation to general nuclear war, with globally catastrophic consequences.

The nuclear threat is certainly out of proportion with the defensive capabilities of a non-nuclear nation. The Swedish government therefore views working for nuclear disarmament as the main defense of a nonallied, non-nuclear nation against this particular threat. Accordingly, nuclear disarmament, and generally the reduction of the role of nuclear weapons in military policy, has traditionally been salient in Swedish arms control policy. An increased nuclear threshold in Europe, by the establishment of a conventional balance on as low a level as possible, would be seen as a very positive development. Still, it is realized that nuclear weapons will be part of great power arsenals for the foreseeable future.[19] This makes it all the more imperative to avoid any war involving the superpowers or the European powers.

Nuclear weapons are commonly assumed to contribute to the deterrence of war. Occasional statements by various Swedish officials notwithstanding, I am certain that this is an accepted view in Swedish decision-making circles too. It is, for example, implicit in the 1984 Defence Committee report.[20] But in the long term, nuclear deterrence is a very shaky foundation for peace, and the risks involved are intolerable. To find ways of establishing more stable foundations for peace, dialogue and a will to accept necessary compromise are evident preconditions. While responsibility for keeping dialogue open undoubtedly rests mainly with the superpowers and their allies, the neutrals can volunteer their good offices and generally try to further the process by whatever means available. A condition for success in such endeavors is, of course, that they be based on a realistic appreciation and understanding of political and strategic realities and perceptions. Another condition is the maintenance of good relations with both East and West.

Unfailing support for the long-term goal of international justice, and for measures to maintain and to strengthen international law, can help reduce the risk of a third world conflict spreading to Europe. The neutrals may be freer to act for the furtherance of these goals, as they are not bound by any obvious ties of bloc solidarity. Besides, it is very much a direct interest of theirs that the principles of international law be respected; territorial integrity, noninterference in internal affairs, and national self-determination are but a few aspects of international law which are of particular significance for small nations, such as the European neutrals.

The neutral and nonaligned nations of Europe have in some instances been able to suggest compromise solutions in international forums. An example of fruitful cooperation in this respect is the CSCE process, particularly at the Madrid meeting. This has not excluded differences of opinion on a number of issues within the N+N group (neutral and nonaligned states), perhaps as large as those between the main alliances. Sweden sometimes finds it easier to arrive at common points of view with its Nordic neighbors than with its neutral fellows. As an example of regular and often very successful cooperation between the Nordic countries one might point at their activities in the United Nations.

Defense cooperation between the European neutrals, even if limited to the four north European ones, is, I believe, not a viable proposition. Even the prospects for joint weapons development or production are rather dim. Military geography, threat perceptions, political factors, and even tradition argue against it. There again, I believe that, despite different security policy choices, the Nordic countries experience less difficulty in finding possibilities of cooperation.

VIEWS OF WAR AND NEUTRALITY:
A PARTIAL CONCLUSION

The perception of neutrality and its viability is determined partly by political beliefs, partly by assumptions about the origin and character of future war, and partly by the relative weight accorded to fighting a war and preventing it; it is also influenced by opinions as to the best way of preventing war in the short and the long term.

As noted above, prevention of war is accorded a prominent place in Swedish security policy, and for Sweden a policy of neutrality is thought to be the policy which best serves this goal. In fact, I believe that its presumed contribution toward the prevention of war could be regarded as sufficient justification of the Swedish policy of neutrality, irrespective of the likelihood of neutrality being a viable proposition if war comes. Besides, neutrality and nonparticipation in great power alliances have become very deeply ingrained in Swedish thinking and public opinion.

Questions measuring support for the policy of neutrality are not normally part of the Psychological Defence Questionnaire. However, when in the autumn of 1986 such a question was included, a solid majority of 93 percent agreed with the policy, while only 2 percent disagreed and another 2 percent were of divided opinion. A similar trend was observed in the results of a private opinion poll taken in 1983, when debate on security policy had intensified in the wake of the submarine incursions. Confronted with the proposition, "We can never get rid of the submarines without American help. Sweden should join NATO," a mere 3 percent declared themselves in agreement. Even among the Conservatives, who generally show the most understanding of American and NATO policy, only 6 percent agreed, while 2 percent of the Social Democrats were in favor. In this connection it is interesting to take note of figure 1.7, which shows that a sizable majority of Swedes believe that the prospects of Sweden managing to stay out of a major European war are slight.[21]

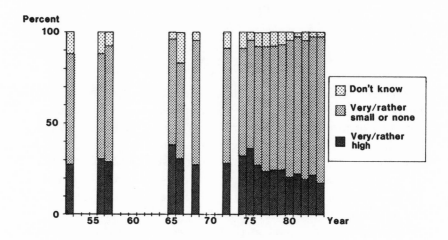

Figure 1.7 Attitudes on Prospects for Neutrality

If it is assumed that a future war can only be a struggle for life or death, arising because one of the superpowers, sensing a threat to the very core of its existence, judged war to be the lesser evil, then one can only conclude that neutrality goes down the drain with everything else. But is such a struggle for life or death conceivable, in the sense of life being a possible outcome? Anyway, to assume that this is the only, or even the most likely, form for war, and to plan accordingly, would be to invite disaster.

A war begun on political grounds, and not as a desperate act inspired by a sense of mortal threat to the existence of a nation, could also deteriorate into a struggle for life or death. In a war fought for political goals conduct would be subordinated to political considerations.[22] In such a war neutrality might be a viable proposition, and the possibility of the neutrals playing a constructive role in such circumstances should not be ruled out offhand.

The Defence Committee report of 1985 describes still another kind of war, one that is not necessarily intended.

It is unlikely that a war between the power blocs in Europe would come about as a result of premeditated decisions to attack in the hope of winning a victory in Europe as a whole. It is more likely that a war would start as a result of a process—with its origins either in or outside Europe—during which successive decisions were made without sufficient information and in a state of uncertainty as to the opposing side's reactions. Furthermore, the process might be exacerbated by circumstances over which the superpowers did not have full control.[23]

The Defence Committee also found it probable that both sides would try to conduct a war at a conventional level, at least initially. The purpose of such restraint, the Committee suggests, would be to allow for negotiations on a cease-fire and political settlement before total devastation has taken place. Again, in this kind of war neutrality may very well be possible. Furthermore, through diligent diplomacy, and by acting as military buffers between the blocs, neutrals might be able to hamper the spread of conflict to, or within, their region. Here, too, the neutrals could find a role as mediators.

CHAPTER TWO

FINLAND

Pauli O. Järvenpää

Finland's attempt to establish a useful and comfortable position as a neutral country in northern Europe in the twentieth century is the story of a struggle.[1] Even in the face of its apparent success at creating a neutral niche in European diplomacy, Finland continues to search for answers to a set of complex questions about its role in regional and European-wide security calculations. That search creates in its wake four broad questions that serve as the framework for any successful exploration of Finland's options.

1. In what kind of a political-security environment does Finland have to conduct its policies? What are the country's historical and geographic legacies that affect its current policies?

2. What kind of security policies has Finland used to cope with its environment, and what are the future defense trends for a small neutral such as Finland?

3. What has been the experience, more generally, of neutrals in the European context, and has it been more or less similar to Finland's? Have recent changes complicated or opened up opportunities for the neutrals?

4. What is the future for Finland's policies, and for those of other neutrals as well, in the Nordic area and Europe generally?

With answers to some of these questions, however elusive they may be, it may be possible to analyze Finland's role in the East-West context.

FINLAND'S DEFENSE POLICY

Framework for National Defense

Finland's defense policy can be summarized by the following propositions. First, the international milieu in which Finland is situated can still be described, in relative terms, as "the quiet corner of Europe."[2] Second, if a crisis breaks out in northern Europe, any aggressor would be unlikely to use more than a small fraction of its total military resources against Finland. Third, the task of Finnish defense is to act, in the most likely case, as a deterrent against such an eventuality and, in the worst case, to make it unattractive to use Finland's territory for hostile purposes. And, finally, the forces Finland can muster would enable the country to perform the tasks required.

Each point is important for carrying out a successful policy of neutrality. The first point proposes that the Nordic area, and the far north in general, has constituted a zone of relative stability and peace since World War II—a barrier in the confrontation between the military alliances. This state of affairs has been the result of cautious and restrictive policies on all sides. On one hand Norway and Denmark have accepted self-imposed restrictions on their NATO commitments, while the Soviet Union has kept the number of its ground forces stationed in the area to a minimum. Sweden and Finland, through their national efforts of deterrence and reassurance, have provided a buffer between the military alliances.[3]

However, there is no escape from the fact that the Nordic region, particularly its adjacent waters, has grown in military significance. Some observers have gone as far as stating that NATO's northern flank no longer is a "flank" but is a "center" of strategic importance: the main strategic axis no longer runs from London or Paris to Berlin and from there on to Moscow; it directly links Washington to Moscow. That means, for the Nordic region, that in terms of the central balance—the balance of strategic nuclear weapons between the superpowers—it is the far north rather than central Europe that will be the main stage of superpower competition and potential confrontation.[4]

The picture of the Nordic area as a center rather than as a flank might be slightly overdrawn, but it is undeniable that although the military situation in the far north has been stable, it has been far from static. There has been the buildup in the Kola peninsula, largely as a result of a shift in maritime strategies by the Soviet Union. The Reagan administration's plan to increase the size of the U.S. Navy to 600 ships, with its accompanying new "Maritime Strategy," will also have repercussions in northern waters.[5] And it is in the north that the new technologies—intelligence acquisition, ASW warfare, cruise missiles, to name just a few—have had some of their widest applications.

Therefore the European north no longer is a backwoods of the European security landscape but is, rather, an area that is intimately connected to the military situation at large. In the 1960s it was still possible to separate, for security calculations, the north from central Europe. But in the 1980s and beyond, the northern areas and central Europe can be likened to a pair of Siamese twins: if security is disturbed in one area, the effects of disturbance are readily felt in the other.

The potential threat in the Nordic region will, therefore, most likely originate not inside the region but as a result of bloc-to-bloc competition. A starting point for Finland's military planning, under these circumstances, is that there are no strategically attractive targets within the country's national territory.[6] Finland does not possess any strategic assets as such; nor has it on its soil any foreign troops, their bases, or nuclear weapons installations that would draw fire. Finnish territory is difficult to traverse and offers therefore no obvious advantages for land operations against a third party—not even in Lapland, although the road network there has improved over the last couple of decades.

As a result, the country is not likely to be a target of an isolated attack, nor would a potential attacker be able to use its most modern equipment or a great number of its forces in the Finnish direction. A military threat against Finland is, in all likelihood, going to be part of a wider conflict between the military alliances.

In such a situation the crucial task for Finland will be to try to remain outside of the conflict as long as possible. Since Finland is not likely to interest any potential attacker, the only military interest it might possibly have lies in its position as an alternative route of advance against a third party.[7] Consequently Finland's primary task is to maintain so strong a defense—that is, to raise the "entrance fee" to the country—as to make it prohibitively costly for any potential aggressor to forcefully violate Finnish territory. The success of the primary goal of Finnish security policy—to keep the country out of all conflicts—depends to a great degree, in crisis situations, on how successful the Finns are in creating and maintaining such a deterrent capability.

International Agreements

Before examining the state of the art of Finnish military defense, it might be useful to briefly discuss some of the international agreements Finland has signed that also contain military elements. One such agreement that has a bearing on Finland's defense posture is the Treaty on the Åaland Islands.[8] According to that treaty, Finland is obliged to protect the neutrality of these islands and to provide for the security of their population under any circumstances, although in peacetime the islands will remain demilitarized. In peacetime Finland may temporarily send armed forces to the islands to maintain order, carry out inspections in the area by one or two light warships, or carry out surveillance from the air.

Another treaty that specifically addresses Finland's defense capability is the Paris Peace Treaty of 1947.[9] In that treaty Finland is authorized armed forces consisting of a total strength that should not exceed 41,900 personnel (army: 34,400; navy: 4,500 and 10,000 tons; and air force: 3,000 and 60 aircraft). Finland is not allowed to possess, construct, or experiment with any atomic weapon, any self-propelled or guided missiles or apparatus connected with their discharge, sea mines and torpedoes of noncontact types, submarines, motor torpedo boats, or specialized types of assault crafts. Nor is Finland allowed to possess or acquire any aircraft designed primarily as bombers with internal bomb-carrying facilities.

However, these restrictions have not had any serious effect on Finland's defense capabilities. In the 1960s, when the ceiling for total peacetime personnel was temporarily broken due to the exceptionally large age groups from the postwar "baby boom" years, the signatory powers to the Paris Treaty were duly notified, and they gave their consent.

Furthermore, when Finland approached the signatory powers in 1963 in order to get the missile clause removed from the treaty, the treaty was reinterpreted to let Finland acquire defensive missiles. When a similar problem arose with influence mines, Finland's wishes were respected and in 1982 the country was allowed to purchase such mines from the main signatories to the Paris Peace Treaty. The reinterpretations have been taken by Finland as a further recognition of its policy of neutrality and as legitimation of Finland's national defense efforts. [10]

The most important treaty with military ramifications is the Treaty of Friendship, Co-operation and Mutual Assistance concluded between Finland and the Soviet Union in 1948. [11] The essential point in that treaty is the strategic interest shared by the two countries: that Finland will in no circumstances serve as a transit route for an attack against the Soviet Union. An examination of articles 1 and 2, the military clauses, indicates that, in terms of concrete military obligations, the treaty is of a unilateral nature. The only military obligation Finland has assumed in signing the treaty is to defend its territorial integrity within its own frontiers by its own forces. If Finland cannot do it by itself, Finland could do it, "if necessary, with the assistance of, or jointly with, the Soviet Union." [12]

The assistance, then, is by no means automatic but is needed only if Finland's own forces will not suffice. The form of aid is also left undefined and, furthermore, the forthcoming of Soviet aid will be subject to mutual consent. It is naturally in Finnish interests to try and keep the situation in Finnish hands as long as possible. [13] In other words, for Finland the treaty states the obvious: as an independent country Finland would in any case fight for its territorial integrity. For the Soviet Union, the treaty removes speculation and a threat of military attack on one of its flanks.

Defense Committees and Appropriations

Practical guidelines for defense decisions have been sketched out by parliamentary defense committees, which since the early 1970s have provided a forum for defense debates and parliamentary input on national defense policy formulation. The first committee prepared its report in 1971, the second committee five years later, and the third in 1981.[14] In 1985 a decision was made not to establish another parliamentary defense committee. Instead, a parliamentary defense commission was set up in April 1985, with a more narrow task of evaluating the Defense Forces's five-year plan for 1987–91. The commission published its report on 26 April 1986.[15]

In terms of concrete defense decisions for the 1980s and beyond, the work of the Third Parliamentary Committee warrants a closer examination. The committee, reporting in 1981, studied the development of Finnish defense capability from both a fifteen-year-long perspective and a five-year term for 1982–86. It might be interesting to take a look at some of the long-term trends seen by the committee.

One of the basic trends the committee noticed was that the available reserves will diminish with the smaller size of age groups— from about 700,000 men (out of the total trained reserve of about 1.1 million) to some 600,000 men in the 1990s.[16] At the same time it also foresaw that it would not be possible—barring a drastic increase in the level of defense spending—to keep the main forces, the bulk of the reserves, equipped to the exacting standards of modern battlefield requirements.

For these reasons the committee chose to separate a part of the main forces for special attention. That part, to be called the Fast Deployment Forces (FDF), was to be given priority over the rest of the reserves. Their strength, including air force and naval combat units, mobile army units, upgraded surveillance, command and control, border, coastal, and necessary logistical formations, could be limited to 250,000 men. They would receive the best equipment and, while flexibly regulated according to the demands of the situation, would be much easier to sustain during prolonged times of increased international tension or crisis situations than a fully mobilized main force.[17]

The practical effects of the committee's recommendations are readily seen today. For the Finnish army the most important development projects are connected with creating and equipping the FDF units. Their mobility, firepower, antitank and antiair capabilities, and combat durability are the main development objectives. Their communications systems must be made reliable in all circumstances. Special attention has been paid to equipping the FDF units earmarked for operations in Lapland under the unique terrain and climatic conditions of that region. Since the FDF units, as all other units, are drawn from the reserves, priority must be given to simple, easy-to-use, and relatively inexpensive defensive weapons systems and auxiliary equipment, so that a reservist can be expected to use them.[18]

In order to bolster Finland's air defense a domestically manufactured communications and radar network has been installed, a squadron of all-weather interceptors to bring the number of interceptor squadrons to three has been purchased, and mobile surface-to-air missile systems that will cover the country's exposed northern border areas have been bought.[19] These systems are also needed for the protection of the most important national targets, such as ports, air bases, and key transportation points. Acquisition of such missile systems still continues.[20]

As to Finland's coastal defense, the committee underlined the importance of repelling naval and amphibious attacks in the Gulf of Finland, the area vital to the defense of the country's capital. In the northern part of the Baltic Sea and in the Åaland Islands priority will be put on naval weapons systems; in the Gulf of Bothnia mobile weapons systems detached from elsewhere will be used, if the need arises.[21] The focus of development of the coastal artillery will shift from fixed to mobile artillery and ground-to-sea missile systems. Three new, domestically built high-speed combat vessel flotillas will replace escort ships and motor gunboats. Specific attention will be devoted to mine warfare capability. Since submarine operations in the Baltic Sea area have increased in scope and intensity, submarine detection systems have been improved to meet that challenge.[22]

The committee estimated that the total price tag for its recommendations would amount to about 17.5 billion Finnish marks (at 1981 prices) for the period 1982–86.[23] That meant a projected annual real growth rate of 3.8 percent, which is a respectable figure by any standards. Growth rates for defense spending have been quite healthy ever since the early 1970s, or throughout the period of the parliamentary defense committees. The average real increase per year during the period 1972–81 covered by the first two committees was 3.7 percent.[24] The projected growth figure for the five-year period 1982–86—3.8 percent—was almost reached, with the actual real growth during that period coming to about 3.7 percent.[25]

Of the defense budget, about 31 percent has in recent years been used for acquisition of military equipment, an increase of about 10 percentage points compared to the spending figures in the late 1960s.[26] In the 1982–86 spending period the air defense's share was about 34 percent, the naval defense's 18 percent, the army's 38 percent, and the remaining 10 percent of the funds were used for purchases not specified by the defense sector.[27] In the next few years, during the "decade of the army," the ground forces will be allocated most funds so that in the late 1980s, the army's share of the military defense budget will be around 50 percent.[28] But at around the middle of the 1990s there will be another shift in spending targets, since by that time Finland's interceptor squadrons will be in dire need of modernization.

During the past decades, roughly a third of Finland's defense procurement have come from domestic sources, a third from the Soviet Union, and yet another third from the Western countries, mainly from Sweden and Great Britain. In the last few years, attempts have been made to purchase as much matériel as possible from domestic suppliers. Some progress has been made toward that goal. For example, in 1987 about 42 percent of all defense procurement came from domestic sources, while the rest of the purchases were split evenly between East and West. It is now estimated that the share of domestic purchases will, by the end of the 1980s, climb up to about 50 percent of all defense procurement.[29] There are, of course, natural limits on how much Finland can produce by itself. For example, an armored personnel carrier and various types of cross-country vehicles are being produced by Finnish firms, but it is highly unlikely that an indigenous tank or an interceptor could be manufactured at a competitive price.[30]

Defense spending's share of the GNP and state spending have remained rather constant since the early 1960s. If one looks at the funds that are available for what in Finland is called "military defense"—i.e., those funds that are directly funneled into defense in the state budget—one gets a stable average figure of about 1.5 percent of the GNP and a little over 5 percent of total state spending.[31] On the other hand, if one factors in various costs, such as the funds available for civil defense, military pensions, the budget for the Border Guards organization (organized under the Ministry of Interior in peacetime), and some other costs, one finds that for all different tasks of military and civilian defense, Finland spends a little over 2 percent of its GNP annually.

The parliamentary defense commission, in its report of 25 April 1986, to a large extent agreed with the long-term trends outlined by the Third Parliamentary Defense Committee. The basic premises of Finland's security policy, as the commission sees them, are still as valid as they were in the early 1980s when the committee published its findings. One new military feature is, however, emphasized in the commission's report. The report states:

> During the past few years, there seems to have been a tendency to limit military conflicts in time in order to use as little military force as possible to prevent the situation from developing into a full-scale war. For that purpose, special troops have been trained. The objective is to attain the military goals by their flexible deployment and scrupulous use. Also, special troops have been trained and equipped for reconnaissance and destruction activities.[32]

In terms of concrete action the commission notes that the basic guidelines for the development of the Defense Forces sketched out by the committee are still valid. But in the commission's view, shortcomings can be found in how the budgetary recommendations by the committee have been followed. "The recommendations for the years 1982–86 were implemented in a way that the salary as well as operation and maintenance appropriations exceeded recommendations but the material procurement appropriations remained lower than proposed."[33] The commission recommended, though without agreement on exact figures, that shortfalls in appropriations should be made up.[34]

REGIONAL AND EUROPEAN SECURITY MEASURES

Finland has, for its own part, attempted to promote relaxation of tension in Europe. In that effort she has had at least three roles. First, she has been an initiator of regional security arrangements. Second, she has strongly supported the establishment of new security relationships in Europe, particularly through the CSCE process. And third, she has been able to provide technical competence in certain fields connected with arms control and disarmament.

Regional Arrangements

The establishment of nuclear weapons-free zones is an arms control measure that represents a geographical approach to solving some of the security problems created by the existence of nuclear weapons. One geographical region where a zonal idea might work is the Nordic area. The Finnish proposal for a nuclear weapons-free zone in northern Europe was first presented in 1963 by the then president of Finland Urho Kekkonen. The idea was reintroduced and broadened in another speech by President Kekkonen in May 1978 in Stockholm.[35]

It is interesting to analyze some of the most central elements of those speeches from today's perspective. According to the Kekkonen speeches, the Finnish objective is to create a separate Nordic treaty arrangement that, as completely as possible, will isolate the Nordic countries from the negative effects of nuclear strategy in general and nuclear weapons technology in particular.

Three points were—and from the Finnish point of view continue to be—central in the Kekkonen proposal. First, the starting point in the Finnish view will have to be the existing security arrangements adopted and maintained by the nations in the Nordic area. Second, it would be, from the Finnish viewpoint, natural and, for the general prospects of establishing a nuclear weapons-free zone in the north, even critical, for the leading nuclear powers to participate from the very beginning in the negotiations establishing the zone.

A third central feature of the Kekkonen proposal has been the notion of negative security guarantees. The argument runs as follows: If the small Nordic countries consciously and irrevocably decide to refrain from developing nuclear weapons to be stationed within their territories under any circumstances, it is only reasonable to expect that they will want to receive guarantees from nuclear powers affirming that these powers will under no conditions use nuclear weapons against states belonging to the zone, nor will they subject the zonal states to nuclear blackmail. The constitution of such a set of unconditional guarantees is seen to provide a cushion against the exertion of pressure by outside powers on any Nordic country, and thus to introduce a stabilizing feature in the Nordic region, particularly in times of increasing international tension.

The Kekkonen proposal met with some rather strong skepticism from the very beginning.[36] The idea of creating a nuclear weapons-free zone in the Nordic area—which at its core is nuclear-free already, totally so in peacetime and conditionally so in time of war—is not without some problems. One has only to ask what the zone's geographical boundaries would be, which of the sea and ocean areas would be included, what would be done to those nuclear weapons that are in the immediate vicinity of the zone—to name just a few possible questions—to conclude that the practical details to be discussed before a nuclear weapons-free zone could ever be established in the Nordic area are indeed complex and not given to easy solutions.

What is most interesting in the Finnish proposal, however, is not so much its present-day practical feasibility, but its long-term security-political goal. Two points need to be emphasized here. First, it would be an important political signal to establish a nuclear weapons-free zone in an area that is not at the center, but very close to the center, of an immensely important strategic area—an area where superpower interests are paramount. Some observers have recently pointed out that the probability for the use of nuclear weapons in the northern sea areas in crisis situations is exceptionally high.[37]

Those areas have, for example, been mentioned as a potential target for what has been termed "warning shots" of nuclear detonations. Furthermore, it has been argued that nuclear war could be the end result of military operations in the north through a process of rapid escalation from conventional to nuclear means of warfare. The far north could experience an "inadvertent nuclear war," as Barry Posen has called it.[38] In the sea areas surrounding northern Scandinavia it will not be easy to make a sharp distinction between conventional operations against conventional targets, on the one hand, and conventional operations against nuclear-capable targets on the other. Military logic might then, in a tense crisis situation, take us down the "slippery slope of escalation," to use the term Thomas Schelling coined long ago, without anybody really wanting to escalate.[39]

Again, what might happen in a real situation is less relevant here than the principle involved. It is impossible, of course, to know what a real crisis would look like. What is undeniable is that it would be impossible to separate for targeting purposes, for example, those submarines that carry submarine-launched ballistic missiles from those submarines that are not so equipped.[40] Also, communications in deep crises would probably be hampered, and decision makers would have to make their crucial decisions in a relatively short period of time under great uncertainty. Inadvertent escalation could indeed be the result of any military operations in the north. Or, as the case might well be, just because that particular region is known to be so highly volatile, the adversaries in crisis situations would tread there with utmost caution.

Whatever the actual outcome of such a crisis in the Nordic area, it is clear that it is in everybody's interests to keep the tension there at the lowest possible level. A nuclear weapons-free zone established in the Nordic area would have a tendency to dampen the incentives for unintended escalation. It would be important to know that in those crucial moments of very severe crises, that geographic area, at least, would not be the cause for escalation.

The nuclear weapons-free zone would also have a peacetime role as a sort of a political "fire break"—an area contractually free of nuclear charges. It is true, of course, that nuclear weapons might be introduced into the proscribed area in a war, and that the nuclear weapons-free zone might in any case be the target of nuclear weapons outside the zone. But such objections do not invalidate the proposed arrangements, since they do not really address the predicament for which the Kekkonen proposal was designed. It was meant, in language derived from the CSCE process, as a "confidence-building measure," which would have the purpose of dampening the incentives to introduce nuclear weapons in situations where it is in everybody's interests not to cross the nuclear threshold, but where the adversaries fear preemption by the other side.[41]

It certainly would be in the interests of nuclear powers in the northern regions of Scandinavia and adjacent waters to help deter a desperate gamble on preemption. And from the viewpoint of the Nordic countries, it would be important to know that the nuclear states would have no reasons to introduce nuclear weapons, their launchers, or their support systems into the Nordic area. A nuclear weapons-free zone would, therefore, capitalize on a common interest shared by all countries of the Nordic area as well as those other states that have security interests in the wider geographical region of northern Europe.

Confidence- and Security-Building Measures

The other area where Finland has played a constructive role is in participation in the Conference on Security and Cooperation in Europe (CSCE). It is also one area where the interests of the European neutrals clearly meet, and where cooperation and contacts between the neutral and nonaligned countries of Europe have been exceptionally high ever since the signing of the Final Act of Helsinki in 1975.

In broad terms confidence-building measures (CBM) have fallen into two general categories. On the one hand, there are those measures that are primarily designed for the purpose of inhibiting political exploitation of military force. On the other hand, one has measures that aim at reducing the danger of surprise attack. One participant has argued: "The Stockholm Conference is different from 'classical' arms control negotiations in that it addresses not the capabilities for war, the number of weapons and troops, but rather the most likely causes of war: flawed judgments or miscalculations stemming from fears of sudden attack and uncertainty about the military intentions of an adversary."[42]

Both of these approaches have been used in the process of designing CBMs, and continued to be as the "second generation of CBMs"—the CSBMs or the confidence- and security-building measures—were being negotiated in the Conference on Confidence- and Security-building Measures and Disarmament in Europe (CDE) that convened in Stockholm. It is far too early to start analyzing the future significance of what the Stockholm Conference has concretely produced. Instead, some general remarks can be made about the significance of the whole CSCE process to small European neutrals like Finland.

First, the CSCE process, including its emphasis on confidence and security-building in the military field, provides an important framework for negotiation and dialogue about concrete measures, which will contribute in the short term toward the construction of a more secure political order in Europe. The results achieved in the CSCE process have proved valuable and tangible. Second, the CSCE process provides opportunities to create, over the long term, more stability and reduced tension in Europe. For a small neutral like Finland that seeks security through primarily political means, this is an important factor. Finland has a permanent interest in strengthening the framework for security and cooperation in Europe. The results achieved in the military field, through the introduction of CBMs and CSBMs, have been modest in scope and significance, but they are a promising beginning.

The third remark has to do, specifically, with the Stockholm Conference and its significance for a small country like Finland. From the viewpoint of small neutrals, Stockholm was a novel forum: all countries in Europe now, for the first time, had the opportunity to participate, in a multilateral forum, in negotiations concentrating solely on questions of military security in Europe. Most of arms control so far has been bilateral, or at least an area of discussion for a few powers. In the CDE, neutrals were able to tread in an area that up to that point had been restricted to the members of military alliances. Military questions, therefore, add a new, significant dimension to the CSCE process for countries like Finland.

Last, but by no means least, it is important to underscore the fact that on all major issues the neutral and nonaligned countries have cooperated to work out a common neutral and nonaligned view. It has not, of course, happened without compromises; but it has happened. And although the national viewpoints can be expected to gain importance in the course of negotiations, when the negotiations might get closer to the core of each country's special security concerns, it also can be expected that in broad terms the cooperation between the European neutral and nonaligned countries that has marked the CSCE process will continue.

Technological Opportunities in Promoting Arms Control

A highly advanced industrial society like Finland can perform some special functions in the area of arms control. Yet a small country must be selective and carefully choose such functions so that it can maximize its impact. In the case of Finland it has been the research on ways to verify chemical agents that has been emphasized in the last few years.[43] As verification has been one of the greatest obstacles to conclusion of a comprehensive ban on chemical weapons, the Ministry for Foreign Affairs established a project on chemical disarmament in 1971. Experimental work, initiated in 1973, has centered on the development of analytical techniques. The goal of the project has been to create a verification capability that would cover all verification needs: verification of nonproduction, destruction of stockpiles, and alleged use.

An additional aspect of the chemical disarmament project is the acquisition of the necessary equipment for verification of chemical agents. The Finnish goal is to create a national chemical weapons control capability, which could be put to international use, if the need arises. Because of the project, Finland has been allowed to be present as observer at the sessions of the Conference on Disarmament in Geneva.

Another case worth mentioning here is seismological verification capability being created in Finland in cooperation with several other countries. In 1976 the Geneva Committee on Disarmament set up an ad hoc group of governmental experts to explore the possibility of international cooperation in the detection and identification of seismic events. Finland has participated in that expert group and is also participating in the network of seismic verification stations that has been built up around the world. Today the network consists of some fifty stations located in different parts of the globe. In Finland, there are two registering multipoint stations, as well as another two single point seismographic stations, that belong to the international network. Such a system of verification of seismological events will be of utmost importance if an agreement is reached on a comprehensive nuclear test ban treaty.[44]

Obviously, it is not a lack of technical know-how, or shortage of resources, that prevents the superpowers from carrying out such arms control related research themselves. The special significance of the Finnish projects stems from the fact that, unlike the superpowers—who would be reluctant to allow the findings of such research to be widely scrutinized—neutral countries, like Finland, can openly share the results of their high-caliber research. Thus they have, in a sense, cornered the market of research on highly specialized areas of verification.

CONCLUSIONS

The most central elements of Finland's security policies and defense have been discussed in the preceding pages. One conclusion that can be drawn from this discussion is that even in the age of sophisticated military technology, all hope is not lost for a successful national defense by small European neutrals. Of course, it is true that if a strong military power were to throw even a modest portion of its military might against a neutral, there would be little hope for the defender. But since neutrals like Finland are unlikely to be objects of isolated attacks, one can assume that even in a worst case scenario a neutral would not need to bear the full brunt of aggression. The object for a small neutral under those circumstances would be to raise the costs to the aggressor to a point where they exceeded the benefits of an attack.

In the Finnish case the large pool of trained reservists, the high mobilization preparedness, the territorial defense system, and the recent actions taken to develop defense capability, as well as the advantages bestowed to the defender by skillful use of the natural features of the country, all indicate that an attack against Finland would not end quickly and that the potential gains would not equal the likely losses.

It is by these measures that Finland will signal to outsiders its determination to repel all violations against its territorial integrity. The fast deployment forces, when they are fully developed, will provide the political leadership with a flexible deterrent instrument. In the next few years, the budgetary process will show how quickly and effectively these goals can be achieved.

On the other hand, a small country like Finland can never rely solely on the deterrent function of its armed forces. Consequently, Finland has, through regional initiatives and in the wider European context, worked for cooperative security solutions in Europe. It has been, and continues to be, in the Finnish national interest to try to promote confidence- and security-building measures for the Nordic area, to cooperate with other countries to create a lasting security framework in Europe, and to offer its technological expertise in arms control questions. There is no doubt that European neutrals will play an important role in working toward those goals.

CHAPTER THREE

AUSTRIA

Heinz Vetschera

ORIGINS OF AUSTRIA'S NEUTRALITY

Compared to other European neutrals, Austria has relatively little experience with neutrality.[1] Both Sweden and Switzerland were able to maintain their neutrality during the world wars. In contrast, Austria was a main actor in World War I. Later, in 1938, Austria was occupied by Germany and subsequently incorporated into the Third Reich. Allied planning in World War II approached Austria in a twofold way. On the one hand, Austria's desire to reestablish independence was recognized; on the other hand, the Allies envisaged a four-part military occupation of Austria by the United Kingdom, the Soviet Union, the United States, and France similar to that of Germany. Liberation from German occupation in 1945 thus meant Allied occupation for Austria. In contrast to Germany, however, Austria could establish a central government and held democratic elections as early as 1945.

This fact proved decisive for Austria. Negotiations for a state treaty with the Allied powers to end occupation were initiated in 1946, along with negotiations for peace treaties between the Allied powers and Italy, Bulgaria, Hungary, Romania, and Finland. These peace treaties would have been concluded in February 1947,[2] but in 1948 the outbreak of the cold war interrupted negotiations for the Austrian State Treaty. Austria could at least avoid Germany's fate of division into an Eastern and a Western state, but both East and

West had reasons to stay in Austria. The West feared that Austria might seek accommodation with the East at the expense of her western orientation. The East feared that Austria might join NATO,[3] being a western democracy. Thus, both sides continued their established military presence in Austria.

New chances emerged with a reorientation of Soviet foreign policy after 1953.[4] At the Berlin Conference in 1954 the Soviet Union proposed a clause to the state treaty prohibiting Austria to join military alliances. Both the Western powers and Austria rejected the idea of imposed neutralization. However, in early 1955 Austria offered to unilaterally declare permanent neutrality in exchange for the Soviet Union's consent to the state treaty. The "package deal" was sealed in Moscow on 15 April 1955 ("Moscow Memorandum"). On 15 May 1955, the state treaty was signed in Vienna between Austria and the four Allied powers, reestablishing Austria's full independence. After the last foreign soldier had left Austria, permanent neutrality was declared in a federal constitutional law on 26 October 1955.

Two facts derive from these origins. On the legal side, Austria's neutrality rests on a unilateral legislative act. It has not been enshrined in the state treaty and is not guaranteed by the signatory powers, nor would it allow interpretation by one or the other signatory power. On the political side, Austria's neutrality is rooted in the cold war. East and West had "neutralized" Austria by exerting joint control, denying each other dominance over Austria. Both sides left Austria only after their security demands had been met. For the West it was Austria's uncompromised commitment to democracy, limiting neutrality to military self-reliance. For the East it was Austria's status of permanent neutrality, preventing her from joining NATO. As mutual mistrust continues between East and West, residual mistrust also remains towards Austria in the form of fear that she could betray her position. Many countries still view Austria as a "gray area," which she was from 1945 to 1955.[5]

AUSTRIA'S POLICY OF NEUTRALITY SINCE 1955

Since the declaration of permanent neutrality in 1955, Austria's policy of neutrality has been shaped by both constant and fluctuating factors.

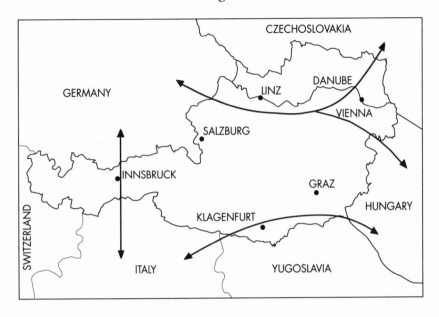

Figure 3.1 Austria's Strategic Significance

Constant Factors in Austria's Policy
of Neutrality

Constant factors in Austria's policy of neutrality derive from the country's strategic location in Europe and from a consensus within Austria on the basic elements of her neutrality.

Strategic Location. Austria is located in an area of traditional strategic significance (fig. 3.1). The valley of the Danube has been a major communication line as well as an invasion route in the east-west direction for more than a millenium. Main routes from central Europe, both to the Italian peninsula and to the Balkans, cross the Alps on Austrian soil.

Austria's strategic significance explains the Allies' reluctance to withdraw from Austria. It also explains the continuing strategic interest of both East and West in Austria. Austria shares borders with NATO member the Federal Republic of Germany and WTO member Czechoslovakia to the north, WTO member Hungary to the east, nonaligned Yugoslavia to the southeast, NATO member Italy to the south, and neutral Switzerland (plus Liechtenstein) to the west.[6] Given the military situation in Europe, the strategic significance of Austria is not likely to decrease and may even increase in the future.

Basic Elements of Policy of Neutrality. The constitutional law on permanent neutrality defines neutrality both as an instrument of security policy—to "secure independence and territorial integrity"—and as an objective for security policy—obliging Austria to "maintain and defend neutrality with all means at its disposal," including military force.

Austria's neutrality is understood as military neutrality, aimed at remaining neutral in any future armed conflict. It does not entail any "neutralism" in ideological, economic, or political issues,[7] or "equidistance." Equidistance implies that one continuously adapts one's own policy toward others' policies. It would make Austria dependent on others and would diametrically contradict Austria's objective of preserving independence. In contrast, Austria's policy of neutrality is aimed at convincing other countries that Austria is willing and able to remain independent and that it will act as a neutral whenever armed conflict breaks out between others.

Second, Austria's neutrality is understood as "armed neutrality." It derives from the legal obligation of any neutral to deny use of its territory to all belligerents.[8] The task to defend neutrality has been expressis verbis enshrined in the constitutional law on neutrality. Austria established armed forces as soon as independence had been regained in 1955. In 1956, general conscription was introduced. A system of Comprehensive National Defense was established in 1961, adding elements of psychological, economic, and civil defense to the military. In 1975 Comprehensive National Defense was enshrined as a mandatory principle in Austria's federal Constitution and order was given to elaborate a defense plan.

The third basic element is Austria's international activities. Austria applied for membership in the United Nations and was admitted in December 1955. Since that time, Austria has been an active member of the UN, participating inter alia as a member of the Security Council from 1973 to 1974. Dr. Kurt Waldheim held the post of general secretary of the UN for two periods (1971–1981).

Austria has also actively participated in peacekeeping operations.[9] In turn, international agencies have established their headquarters in Vienna, making it the "third capital" of the United Nations, besides New York and Geneva. In Austria's perception the status of permanent neutrality thus does not entail passivity in foreign policy but enables it, on the contrary, to conduct an active foreign policy, transcending the neutrals' traditional task of "good services."

The rejection of neutralism in ideological or political issues, the understanding of neutrality as "armed neutrality," and the perception that neutrality offers chances for an active foreign policy form the basic consensus of Austria's policy of neutrality.

Fluctuating Factors in Austria's Policy of Neutrality

Although Austria's policy of neutrality has been based on the constant factors outlined above, fluctuations in policy formulation have also occurred. These fluctuations can be grouped into three historical periods.[10]

1955 to 1970. This period was dominated by the conservative "People's Party." Neutrality policy rested on the concept of "armed neutrality," meaning the establishment and strengthening of Austria's armed forces. These were structured as a cadre/conscript army and resembled small-scale NATO forces. In hindsight, these forces were not tailored for the specific task of defending neutrality, which requires a credible dissuasive posture against any attempt to cross neutral territory. Overall the forces were short of manpower and emphasized armor; they also lacked adequate air cover. Yet they proved adequate for the border crises Austria had to cope with in this period. During the Hungarian uprising in 1956 almost 170,000 Hungarians fled to Austria and there was an inherent danger that fighting between Hungarian and Soviet forces could also spread to Austria. Austrian military and police forces guarded the border and contained the danger. When Soviet and other WP forces invaded Czechoslovakia in 1968, the population did not offer armed resistance to the invasion. Yet, the danger of escalation was inherent, requiring military security operations on Austria's border to Czechoslovakia.[11]

One year earlier, Austria had to cope with a crisis in Italy. There the German-speaking minority conducted a small-scale guerrilla war against Italian security forces because of Italy's delay in granting minority rights. Austria supported the minority's demands politically, but she sealed the borders to cut supplies for the guerrillas from Austrian territory and to facilitate the search for a political solution.

Foreign policy was therefore not neglected during this period as an instrument of security policy. Yet it was perceived as requiring a solid defense policy. However, in the late sixties the credibility of Austria's defense and security policy declined rapidly. One reason for this was the evident gap between the tasks given to the armed forces and their size and capabilities. More important was an overall pacifist climate in Western Europe. In 1970 a people's initiative was launched in Austria to abolish the armed forces and rely on nonviolent resistance.[12] It was supported only by the tiny Communist party and never reached the required votes. The pacifist trend was adopted, however, by the Socialist party, who entered the 1970 elections with the promise to cut military service from nine to six months and won the campaign.

1970 to 1983. The elections initiated the second phase, which was dominated by absolute Socialist majority, led by Federal Chancellor Bruno Kreisky. Defense was deemphasized, while foreign policy was emphasized, as instruments of security policy. One reason for this was the Socialist party's antimilitary tradition, stemming from the Austrian Civil War in 1934, when army and police forces had crushed the socialist workers' militias. Another reason was Kreisky's background as a diplomat who apparently preferred diplomatic over military instruments and tended to neglect strategic issues.[13] Finally, deemphasizing the military corresponded to the election promise to cut military service. In sum, the administration's philosophy at that time read that "a good foreign policy is the best defense policy."

Ironically, cutting military service led to major reform of the armed forces, raising both manpower and prestige. Austria adopted a militia system based on six months' basic training and at least sixty days' refresher training, which provided greater total manpower than the earlier cadre/conscript system. Simultaneously, Austria adopted a strategic concept of area defense, making Austria's defense posture more credible than before. As a consequence, acceptance of armed defense rose to almost 90 percent.

The positive results must not gloss over the fact that essential defense requirements were neglected. Military budgets were kept tight, and the overdue modernization of the air force was delayed indefinitely. Large sums were instead invested in making Vienna the third capital of the United Nations, as though the presence of international organizations would guarantee Austria's security.[14]

Similar developments could be seen elsewhere in Europe, as the overall climate of détente favored foreign policy over defense. However, Austria's foreign policy also shifted from prior patterns, coming closer, on occasion, to the position of the nonaligned movement than to other European neutrals.[15] Yet it would be incorrect to perceive Austria's position as neutralist, since it remained firmly rooted in the West.[16]

1983 to Present. The transition to the third phase came when the Socialist party lost absolute majority in 1983 and formed a coalition government with the Liberal party. Equal emphasis was given to defense and foreign policy in Austria's security policy. The National Defense Plan, which had been long delayed, was finally adopted by the government in 1983 and published in 1985.[17] Modernization of the air force was approved in 1985, introducing supersonic interceptors.[18] In part, the development was due to the influence of the Liberal party, which has consistently supported a strong defense, but it also reflects an overall shift within the Socialists toward a centrist position, which became visible in the conduct of foreign policy as well. After a change in the Liberal party's leadership, the coalition broke and elections were held in November 1986. As no party could gain absolute majority, a coalition government of the Socialist party and the conservative People's party was established in early 1987, which introduced a program of economic austerity. Although this brought severe cuts in the defense budget, Austria has not returned to a "foreign policy first" idea, but security policy continues to be based upon defense and foreign policy on an equal footing.

AUSTRIA'S CURRENT SECURITY POLICY

Austria's current security policy was laid down in the National Defense Plan. It was initiated, together with the federal constitutional law on comprehensive national defense, as early as 1975. Later developments, however, delayed its elaboration and adoption until 1983.

The Present Situation

Austria's Threat Perception. Austria's threat perception differs widely from that of the alliances. Both NATO and the WTO identify each other as the main source of threat. Austria cannot perceive future threats in the same simplistic way, as each of the alliances could be interested in control over Austria for strategic reasons in a future conflict. Austria must identify *what,* rather than *who,* could threaten Austria's security.

Levels of future conflict include that which can occur during "relative peace"—economic or political pressure and terrorism; subversive/revolutionary warfare; conventional war; and limited and unlimited nuclear war. For Austria's defense policy, conventional war forms the main threat. This is true because conventional war is also perceived as a major contingency by both alliances, as reflected in their force structures. Furthermore, in conventional war between the alliances (also between individual states) neutral territory could become a strategic asset to the adversary, and the neutral state would want to deny access, or at least severely limit it. This situation is less important for lower levels of conflict—as, for example, subversive/ revolutionary warfare—but it would be very significant in a full-scale nuclear war.

In a future European conflict, NATO would probably be more able than WTO to break Austria's neutrality on the ground. The WTO has more ground forces than NATO[19] and can divert parts of them more easily. Also, outflanking NATO's defense on the central front via Austria offers the opportunity to threaten NATO's flank in southern Germany. NATO lacks the ground forces for similar options, but probably would need them to hold the central front; there appear

no significant strategic objectives to be achieved by attacking the
WTO via Austrian territory.

In airspace, however, NATO could be the first to violate Austria's
neutrality in a future conflict if it attempted to interdict reinforcements of the WTO by air raids.[20] To avoid the WTO's dense air defense
above the fighting forces, Austrian airspace could be used to outflank
them for attacks into Eastern Europe. Austria thus has to maintain
a credible force posture both on the ground and in the air in order
to maintain her neutrality against both sides.

Singular attacks on Austria appear less likely than they would
be in the context of a major European conflict. Yet one alliance
might use Austria as a "test case" (being still a "gray area") for the
resolve of the other side. Risks of escalation might appear minor
compared to attacking a member of the opposite alliance. In this
case Austria would have to offer resistance against the attempt to
create a fait accompli and must escalate the conflict to the point of
uncalculable risks for the aggressor.

Threatening as these scenarios appear, others may be more
probable. During the past thirty years Austria has been threatened
three times by limited conflicts in her vicinity. As long as détente
and deterrence prevent the outbreak of war between the alliances,
intra-alliance conflicts, or similar forms of low-level conflict, might
be more likely to challenge Austria's security than full-scale European war.

Austria's Strategic Concept. Austria's strategic concept[21] has been
primarily shaped by the desire to preserve neutrality during an armed
conflict between others. Belligerents are most likely to attack a
neutral when they cross neutral territory to save time, undercutting
their adversary's reaction time and catching him by surprise. The
aggressor would need to keep losses as low as possible and would
want to break the neutral's resistance without major deployments of
troops for occupation. Knowing this, Austria has prepared a strategic
concept of area defense. It trades calculated losses of territory against
the aggressor's losses in time and casualties, in order to deny the
surprise effect against the aggressor's main adversary and to raise the
costs to the aggressor. Its strategy is to keep forces intact, posing a
continuing threat against the enemy's rear and flanks, and to force

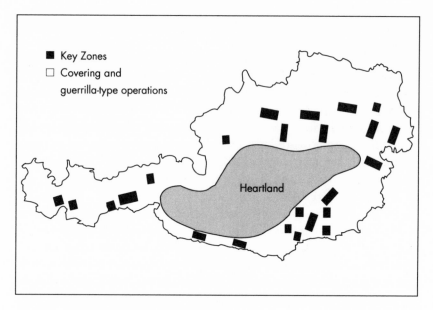

Figure 3.2 Austrian Defense Preparations

the aggressor to saturate Austria with occupation forces, raising the costs beyond the benefits for the aggressor.

Future aggressors would be primarily interested in controlling the main communication lines. Therefore, zones along these lines have been prepared for static defense by fortifications and prepared demolition, which should achieve both delay and attrition of the aggressor (key zones). Other areas would be denied the aggressor by light infantry forces employing guerrilla-type warfare against support, logistics, and communications elements as well as against second-echelon forces (security zones) (fig. 3.2).[22] In addition, security operations should protect borders and other areas of the territory against lower levels of conflict. These operations should prevent escalation and are similar to those conducted during the past crises.

Structure of the Armed Forces. Austria's armed forces have been structured mainly as militia forces.[23] The main force consists of territorial militia forces ("territorial *Landwehr*") to fill the key zones and security zones, organized in regiments, battalions, and companies and structured to their specific tasks. They are reinforced by eight mobile infantry militia brigades (mobile *Landwehr*).

Austria also keeps one mechanized division and one air force division as standing forces (*Bereitschaftstruppe*, or "alert force"), which can cope with minor threats without mobilization of the militia. In case of major threats, these forces should protect mobilization of the militia and reinforce the militia forces in their respective zones.

At present Austria could mobilize more than 200,000 troops, and the National Defense Plan envisages a mobilizable force of 300,000 by the mid-nineties.

Existing Deficiencies

The "Manpower Gap." Trends indicate that Austria may not reach the force level of 300,000 troops in time. First, the birthrate has declined. Second, a law from 1974 allows nonmilitary service for conscientious objectors (COs), reducing annual recruitment by more than 4,000. Finally, too many recruits are absorbed within the military in peacetime functions, for example, as drivers, cooks, or office clerks. They serve a continuous eight-month term but are exempted from future assignments with the militia.[24]

These problems are not insurmountable. COs appear to be recognized and exempted from military service too easily, which may be changed even within the existing legal framework. Also, the army's peacetime work force may be reduced in favor of training a higher percentage for the militia. These measures would suffice to close the manpower gap. More important, however, are the budgetary constraints that have prevented provision of proper equipment for 300,000 troops in time. Thus, a revised force structure (*Heeresgliederung* 87) was introduced in 1987, streamlining the forces toward a mobilizable force of 260,000 troops by the mid-nineties, without, however, abandoning the final objective of a mobilizable force of more than 300,000.[25]

The "Missile Gap." Austria's missile gap[26] stems from a misinterpretation of an arms limitation clause contained in the state treaty as well as the peace treaties of 1947.[27] The clause prohibits "special weapons" such as atomic weapons, submarines, and manned torpedoes, as well as "self-propelled or guided missiles." The clause was introduced into the Italian peace treaty by the United Kingdom to deny access to strategic weapons systems, mainly to prevent circumvention of future arms limitations in a German peace treaty via Germany's former allies.[28] It found its way into the other treaties, including the Austrian State Treaty. The term "missiles," however, has a specific meaning in the terminology of the treaties, making them "special weapons." An addendum to all treaties regarding war material enumerates "rockets, self-propelled and guided missiles and (self-propelled and guided) projectiles" as separate categories. Historic evidence shows that the term "missiles" refers only to weapons with a range of more than 30 kilometers,[29] whereas short-range weapons constitute "self-propelled and guided projectiles" in this terminology and would not be affected by the prohibition clause.

After having already acquired some hundred guided antitank weapons without drawing protests from the signatory powers in the early sixties, Austria then began asking for abrogation or reinterpretation of the arms limitation clause. All attempts to change the treaty were, however, turned down by the Soviet Union.

Area defense requires adequate means to counter enemy tanks, attack planes, and helicopters. The credibility of Austria's defense will rest on the will to close the missile gap. There should be no legal obstacles to accomplishing this, since only short-range "projectiles" (in the language of the treaty) would be needed. Nevertheless, it took a considerable effort to prepare the ground in Austria and with the signatory powers in order to prevent any misinterpretation of Austria's intention to close the "missile gap."[30] According to statements of the federal minister for defense, first trials with antiaircraft missiles will begin by late 1988.

The "Civilian Protection Gap." The main communication lines across Austria are densely populated. Preparing key zones and security zones in these areas should be accompanied by measures to reduce collateral damage among the civilian population, as both the National Defense Plan[31] and the First Additional Protocol to the Geneva Conventions[32] demand adequate civilian protection. Austria's present practice is evidently inadequate: shelters are not available in sufficient numbers. One reason for this may be that earlier attempts to achieve nuclearproof protection were perceived as too costly and were therefore abandoned. A more important factor is Austria's federal structure, which requires state legislation of civilian protection programs. This has, until now, prevented Austria-wide legislation for civilian protection.

The Chernobyl incident may have initiated some change. Civil defense organizations—for example the Red Cross and the militia fire brigades—have achieved a high standard in training and equipment and also high prestige, especially in rural areas. They can cope with natural disasters as well as the effects of armed conflict. Stockpiling and shelter construction until now have been left mainly to private initiative, but today are increasingly perceived as a public task.

AUSTRIA'S NEUTRALITY AND THE FUTURE OF EUROPEAN SECURITY

Developments in the military as well as in the political sphere indicate that European security is likely to become more complex, challenging Austria's policy of neutrality.

Developments in the Military Framework

Developments in Military Technology. Most spectacular developments in military technology have emerged in the area of "exotic" weapons systems, as for example in the SDI project. Yet technological revolution also takes place in less spectacular areas, both as a spin-off from "exotic" developments and independently, and with no lesser impact on the military. For example, new materials have been introduced in armor, rendering many antitank weapons obsolete in the future. Computer technology allows target acquisition and guidance systems for precision-guided munitions (PGM), which may induce far-reaching changes in structure and doctrine of the armed forces.[33]

The impact of these developments on Austria's security cannot be easily estimated yet. Such calculations are still far off for long-term projects, like SDI. Even the initiators offer differing views of the final outcome. Moreover, any developments on the level of strategic nuclear armaments, or defenses against them, appear to be beyond Austria's reach, limiting their immediate consequences on Austria's defense policy whatever the results of SDI might be.

Austria is more concerned with developments immediately influencing her defense posture. For example, Austria has reacted to new armor by developing kinetic-energy rounds which should be able to penetrate any armor Austria's defense would encounter in the foreseeable future. PGMs may gain relevance for Austria, as they could improve Austria's defense capabilities, but also because they could change structures and doctrines of the armed forces in Austria's neighborhood.

Changes in Doctrine: The Trend Toward Conventionalization. A major role for PGMs is seen to be the replacement of nuclear weapons in NATO's doctrine of flexible response.[34] Conventionalization, however, has already been undertaken by the WTO, apparently in order to undercut NATO's nuclear threshold. The revival of shock forces in the Operative Manoeuver Groups (OMG) is further evidence of this trend. NATO apparently followed suit. First, there have been proposals to raise the nuclear threshold by increasing conventional firepower. Then, new battle doctrines stress higher flexibility in counterattacks, extending the battlefield into the rear of the enemy and destroying follow-on-forces.[35] Finally, the INF agreement eliminates a complete category of nuclear weapons, adding to the trend toward reducing reliance on nuclear weapons and emphasizing conventional options.

Consequences of conventionalization are not easily calculated for Austria. On the surface, reducing the danger of nuclear escalation would appear positive, as collateral damage and nuclear fallout would not respect Austria's neutral status, even if the belligerents would do so. However, given Austria's main interest in maintaining stability and peace in Europe, raising the nuclear threshold may appear less positive, if it means raising the specter of conventional war.

Furthermore, doctrinal developments accompanying conventionalization indicate that space will become an increasingly important factor in military calculations. The value of neutral territory and airspace as strategic assets is also raised, increasing the threat for neutrals, particularly Austria with its geographic position.

Changes in Doctrine: The Trend Toward Unconventional Warfare. Both alliances increase their capabilities for unconventional warfare by establishing special forces.[36] Unconventional warfare appears to be designed mainly for extra-European theaters, but it cannot be excluded for Europe. Actions of covert warfare may occur even in peacetime, hidden in an overall scenario of domestic terrorism.[37] They are likely to occur in crises preceding the outbreak of higher levels of conflict.

The latter aspect gains specific significance for Austria. Her defense capability rests on undisturbed mobilization of the militia forces. Subversive acts preceding an armed attack on Austria could delay mobilization with decisive results. Present planning envisages shielding mobilization by early deployment of mechanized forces in threatened areas. Future developments may require an entirely different shield for mobilization against new types of attack.

Developments in the Political Framework

Interalliance Relations and Arms Control. Relations between the alliances have been focused primarily on arms control. Austria does not participate in bloc-to-bloc negotiations, as for example the MBFR/MURFAAMCE talks.[38] She cannot influence their outcome, although it could affect her security by reshaping the strategic environment. Austria's security interest would be served best by lowering the risk of military confrontation in Europe, and encouraging a balance of forces at the lowest level possible.

Furthermore, stability would be enhanced by reducing the alliances' offensive elements. Doctrines and structures geared to offense create inherent pressures for preemption, especially in times of crisis. European security will be served best by preventing escalation at its roots. Unfortunately, discussions are too often devoted to preventing escalation from conventional to nuclear war, as for example in proposals for nuclear weapons-free corridors or zones,[39] or in the no-first-use idea. They ignore the catastrophic consequences even conventional war would entail for Europe, and contribute little to European or Austrian security.

Austria's strictly defensive attitude toward area defense has been advocated by some West European pacifist groups as a model on which to unilaterally restructure NATO's defense.

Such suggestions ignore the difference between Austria's and NATO's strategic objectives. For Austria it may suffice to delay the aggressor's advance for some days in order to dissuade him from aggression at all.[40] In NATO's case, offering resistance in order to delay the aggressor's advance for some days may provide no dissuasion at all. "Transarming" NATO toward area defense is likely to produce instability in Europe and runs contrary to Austria's security interests, even if it does quote the "Austrian model force structure."[41]

Austria has adopted a limited and more realistic approach to arms control. First, Austria has participated in only one major forum, namely the Conference on Confidence- and Security-Building Measures and Disarmament in Europe (CDE) within the framework of the Conference on Security and Cooperation in Europe (CSCE). There, the scope was limited to Confidence- and Security-Building Measures (CSBMs) including information, communication, and measures of constraint.[42]

The CDE approach to arms control corresponds to a view widely shared in Austria—that the military situation is the consequence rather than the cause of East-West tensions. It appears that substantial CSBMs would deal with the causes, rather than the symptoms, of mistrust. Clearly, simple formulas cannot be applied to the varied and many structures of armed forces found in Europe. Only after regulative measures for military conduct are applied will meaningful arms limitations, or even reductions, be achieved. In sum, CSBMs offer more opportunities for European security at the present stage than any farfetched ideas of European disarmament or measures narrowly designed for small sectors of armaments, as for example nuclear weapons.

Intra-Alliance Relations: The Future of Stability. In the overall framework of European security, the alliances are likely to remain dominant factors. Yet internal stability and cohesion of the alliances have already shown signs of decline that are likely to continue.

The alliances view each other as the main source of threat and fear the other's strength, but instabilities in the other alliance are not always to their comfort. First, there is the specter of contagion, as occurred when pacifist ideas originating in the West did not stop at the intra-German border, causing problems in the East too. Secondly, instabilities within the alliances bear the danger of uncontrollable escalation and could spill over into interalliance conflicts. This is even more dangerous for Austria. Events in the Hungarian and Czechoslovak crises did not directly threaten NATO but contained enough threat potential to worry Austria.

Austria's security thus requires stability not only between, but also within, the alliances.[43] This must not be misinterpreted to mean that Austria would refrain from advocating more freedom and human rights. Such a view would, first of all, contradict Austria's identity as a pluralistic Western democracy. It would contradict the objective of security too. Suppressing human rights may stabilize the situation superficially, but it is likely to create the potential for even worse unrest and instability. Advocating human rights contributes to long-term stability and security in Europe.[44]

European Security and the Framework of the CSCE

Austria has actively participated in the CSCE since preparations for it began in 1972 for several reasons. First, the beginning of the CSCE process fell into the "foreign policy first" period of Austria's security policy, offering the chance for an active role. Second, the CSCE is the only European-wide forum on security that allows the neutrals' participation. Third, the main subjects of the CSCE correspond to Austria's main concerns in European security, namely improving the overall security situation, economic relations, and promoting human contacts and human rights in Europe.

Finally, some approaches within the CSCE reflect Austria's views, as, for example, in arms control. The CSCE approach via CBMs and CSBMs appears more promising than the arms reduction idea of MURFAAMCE, because it avoids the problem of comparing entirely different force structures. Furthermore, the CSCE approach to human rights—enshrined in the "decalogue" of principles—reflects the view that respect for human rights enhances security.

Chances and Limits for the European Neutrals

By tradition, neutrality has been understood as self-reliance, excluding any idea of forging a separate bloc. These limits are still valid.[45] Yet within the CSCE a chance emerged for the neutrals to present their security interests vis-à-vis the East and the West. This is especially true for the first basket, where cooperation among the neutrals and with nonaligned countries has shaped a separate Neutral and Nonaligned (N+N) group.[46]

The experience of close cooperation in the CSCE process has apparently led to increased mutual interest among the neutrals too. Limited cooperation among the neutrals has always existed; it is compatible with neutrality as long as it does not oblige to mutual assistance in armed conflicts. Arms transfers have been a regular pattern,[47] as have been exchanges of military personnel. But it appears that there may be more potential for cooperation. Austria has deduced from the CSCE experience that each neutral has gained special experience in certain areas, and that each could learn from the others. In turn, such cooperation could lead to a division of labor among the neutrals for their mutual benefit. The Austrian ministry of defense has initiated annual seminars for defense and security experts in order to increase the exchange of information and to facilitate future cooperation in the area of security policy.

CONCLUSIONS

Austria may be viewed as the most typical European neutral because of her strategic location and her history. Neutrality emerged in the framework of East-West confrontations. Each side has since had a continuing interest in preventing the other's control over Austria, perpetuating Austria's role in the cold war as a "gray area." Austria's strict neutrality in military and security matters, and her uncompromised adherence to Western democracy, can be perceived by both sides as a split identity, leading to residual mistrust against Austria.

Mistrust may have been reinforced by several fluctuations in Austria's policy of neutrality in the past decades. Compared to the fluctuations that the overall framework of East-West relations has undergone in the same period, however, Austria's policy of neutrality appears remarkably stable.

Austria's security is linked to European security in a twofold way. On the one hand, Austria's security depends on European security. Tensions in Europe may mean threat to Austria, even if they are not yet perceived as such by the alliances. On the other hand, any major threat to Austria's security will also threaten European security, since Austria is located in an area of high strategic significance, which cannot be simply neglected.

European security has become increasingly complex. First there is the overall framework, where confrontation and détente alternate. Furthermore, there are tendencies within the societies and alliances (both Eastern and Western) toward less stability and cohesion. The East-West confrontation still forms the source for the most threatening armed conflict that Europe, and Austria, would encounter. Yet other types of conflicts may grow to importance in the future, threatening security either by their potential to trigger an East-West confrontation, or by their own destructive potential.

Maintaining neutrality may become increasingly difficult under these circumstances. Developments in the military field may make it even more difficult to maintain a credible defense posture in the future. New types of conflict could challenge the concept of armed neutrality, as, for example, low-intensity conflicts, civil strife, or terrorism, which cannot be contained in the same way as regular military operations. Finally, Austria's foreign policy will be challenged as it attempts to defuse sources of future tensions and crises that are more complex than the present ones. In sum, security policy has never been an easy task for small neutral states, which are bound to self-reliance in security policy. As European security grows more complex, security policy will become increasingly complicated for Austria too.

CHAPTER FOUR

SWITZERLAND

Laurent F. Carrel

INTRODUCTION

The central question for Switzerland is whether neutrality can guarantee security, and if so, is it a valuable instrument for the prevention of war. There are plenty of historical examples that suggest we should answer this question in a cautious and sober way. If, in today's environment, attempts are made to escape from the NATO commitment by pointing to Switzerland as an example, then we should keep a minute's silence in commemoration of the neutral countries ravaged in recent history. For example, what did neutrality bring Belgium during World War I? What did it bring the Netherlands, Denmark, and Norway during World War II? What were the fatal consequences of the fact that the Hungarians wished to be neutral in 1956, as well as the Czechs in 1968? And is it not true that for nine years a superpower has been waging war against Afghanistan, which, before losing its independence, was a strictly uncommitted country. It seems that without additional security measures, a policy of neutrality has hardly ever successfully protected a country from war or occupation. Even if Switzerland were a case in point of a hitherto successful policy of neutrality, it would be wrong to view the policy as unproblematic or even as an ideal condition. Neutrality carries with it a risk of being alone at the decisive moment, and of being divided up according to the interests of the big powers. In general, it is only during times of crisis or war that neutral states are tested. As history shows, the outcome of such tests is not predictable.

Neutrality cannot be equated with security policy or foreign policy. According to the Hague Convention (V) of 1907, neutrality primarily consists of the obligation not to take part in a war between other states. With that obligation, a number of rights and duties arise for the neutral state. These were extended for the permanent neutrality of Switzerland, which was expressly recognized as being in the interest of Europe as a whole by the great powers of 1815. Among other things, the permanently neutral state is under the additional obligation to defend its neutrality and independence unilaterally. Thus, Switzerland's permanent neutrality is only conceivable as an armed neutrality.

It would be a tragic illusion to believe that security could be guaranteed by neutrality alone. The basic forty-page strategic document on Swiss security policy deals with neutrality on less than a quarter of a page. All it says about this topic is that Switzerland makes its first contribution to international peace and security by keeping its own house in order, and its second contribution through its straightforward policies toward its neighbors and the whole community of nations. A policy of neutrality, carried out without any compromise, shows a clear determination to defend one's independence against all powers. Switzerland will not take actions in peacetime that would be incompatible with its obligations to protect its neutrality.

The strategic document deals with other topics in much greater detail: army and armament, civil defense, the economy and national finances, information, psychological defense and state security, infrastructure for armed resistance and survival, leadership, and general defense. This shows that the policy of neutrality ranks prominently within the security policy of Switzerland, but it does not prevail over all other aspects.

Neutrality cannot be equated with foreign policy. Although neutrality is a mainstay of Swiss foreign policy, it is only concerned with a sector of foreign relations, namely the attempt to stay away from conflict between other states and not to seek security in a pact. Within the scope of Swiss security policy, neutrality and foreign policy are related insofar as the Swiss foreign policy has the task of (1) representing the political principle of armed neutrality abroad and employing it with the purpose of strengthening Swiss dissuasive strategy, and (2) in case of war, contributing to Swiss defense goals. However, the security policy allows for a much broader interpretation of the mission of Swiss foreign policy.[1]

In sum, we can say that neutrality cannot replace an active security and foreign policy (which will have to become even more active in the future). Even less can neutrality be an alternative to defense efforts.

SWISS NEUTRALITY

One cannot talk about a policy of neutrality today without first pointing out the tension existing between the duties of permanent neutrality and the interdependence of today's world. For a highly developed but small industrial country, which has practically no raw materials of its own, limited energy resources, and a small domestic market without access to the sea, this tension is particularly distinct. For one thing, it is the neutral state's duty to avoid being bound to potential parties in a conflict. It must not put itself in a position in time of peace which will prevent it from remaining neutral when a conflict breaks out. In conducting its policy of neutrality, it must do everything it can to avoid being drawn into a war; it must not do anything which might involve it in such a conflict. According to international law, the restriction of a state's sovereignty due to neutrality must be interpreted restrictively. A permanently neutral state still has a considerable degree of freedom of action in conducting its foreign policy, in order to maintain its independence and further peace and international security. Abstinence in foreign policy can hardly be the neutral's answer to the growing interdependence and mutual involvement of nations.

It is by no means a cliché to view humanity as a "community of fate" in which no one can ignore the big problems—the conflicts between East and West, North and South, hunger, supply of energy and raw materials, unemployment, etc. A policy of neutrality, as Switzerland understands it, includes international cooperation and solidarity. It is the neutral state's chance to render a number of services, for which other nations may be less suited. Among these are diplomatic mediation, humanitarian aid, development of international arbitration, cooperation in international organizations, and participation in supervisory commissions. Furthermore, a policy of international presence also includes development aid and other activities expressing solidarity across the borders. The Swiss policy of neutrality is guided by the principles of "neutrality and solidarity." Principal foreign policy activities in support of this effort thus include:

1. participation in the efforts of the main powers aimed at peace and détente (e.g., in the so-called N+N group at the CSCE conferences);

2. participation in international conferences and organizations of mainly economic, technical, scientific, and cultural nature (e.g., UN specialized agencies, OECD, GATT, Council of Europe, International Energy Agency);

3. a policy of availability, in case good offices in the technical or organizational area should be required for mediation (e.g., the Iran hostage affair), protection for states that have broken off diplomatic relations, granting of hospitality to international organizations and conferences, humanitarian activities in case of war or catastrophe, and verification efforts.

The Swiss policy of neutrality has undergone considerable change over time. After being interpreted restrictively, it was broadened and given new meaning after World War II. Today it is far more than a means of state policy for the maintenance of national independence. It has become an instrument applied (with the limited possibilities of a small state) toward the promotion of world peace, the reduction of tensions, and the furtherance of human rights and other nations' independence.

There are many factors limiting the possible activities of a small state. In spite of its many strong points a small country has very little influence on the actions of European and world powers. High hopes regarding the possible influence of small neutral states' foreign policy usually have been smashed on the hard ground of reality. Even if a more active policy of neutrality were an absolute necessity, it would have to be a policy of small and modest steps. In contrast to this, the security policy of an armed and neutral Switzerland, which can convince all foreign powers that every trespass on its territory will cost a high entrance price, may be the most valuable contribution to strategic stability in Europe, and may therefore be understood as an act of European solidarity. In order for such a policy of neutrality to be feasible and credible, it must be founded on a readiness to defend it if necessary. There is no credible neutrality—and no credible Swiss policy of peace—without an army. The conviction expressed by the federal council in its report on the security policy of Switzerland in 1973—"that we can successfully undertake our peacekeeping efforts in the future only if at the same time we can ensure our own security in a credible way"—is still valid today, as is the conclusion that the security policy of a country is credible "if a realistic evaluation of threats and a sober assessment of its own possibilities lead to the implementation of a concept (of general defense) capable of inspiring confidence at home and respect abroad."[2]

EUROPE'S SECURITY AND SWITZERLAND'S DESTINY

Forty years after the end of World War II Europe is not safe, and cannot defend itself, on its own. Europe still depends on the strong protection of the United States and on that country's ability and determination to guarantee Europe's interests in addition to its own. Militarily speaking, Europe needs the American power of deterrence more than ever—and not only the nuclear, but also the conventional one. These facts would hardly be worth mentioning if there were not a strong tendency in Europe to make them disappear behind inflated illusions. Many intellectual concepts, which can easily do without the security guarantee from the United States and which flirt with the free interplay of forces between East and West, give us the tempting illusion of security. All efforts to neutralize the Federal Republic of Germany belong to this line of thought. Even if the FRG, as in the case of France and Spain, were detached from the military but not the political organization of NATO, the alliance would disintegrate, having lost its most important European partner and foundation. At the same time the United States would lose its military foothold in Europe, and the prime Soviet goal of decoupling Western Europe from the United States would be reached. The resulting vacuum would inevitably be used to its advantage by Europe's strongest power, which would utilize its strategic superiority to attain political dominance. The ultimate aim could be to close the gap in central Europe's line of neutral countries by means of a neutral West German bar, thereby extending the USSR's buffer zone reaching from Finland to Yugoslavia. Breaking up the U.S.-European security bond would subject Europe to extreme pressure, during which the security of states like Austria and Switzerland would certainly deteriorate.

The effects of this change are clear. First, Europe's security and the future of its security policy would considerably influence the destiny of Switzerland's security policy. In its own interest Switzerland must favor the continuation of deterrence and of an alliance in security matters between the United States and Western Europe. The notion of independent "Western European Security politics between the blocs" is an illusion, just like the idea that neutrality can be a guarantee for peace. This is not to say that there is no room for Europe to play an essential role within the Western security policy framework—but it will always be dependent upon the American position. Second, there is no reason at all why Switzerland, the small state which is often accused of behaving like a schoolmaster or of considering itself Europe's "model boy," should promote the export of its neutrality concept to other countries.

Even if Switzerland forms an indissoluble part of Europe in many respects—intellectually, culturally, economically, and politically—the country has always tried to reserve a little scope, however small, for its own decisions. While opening up toward the outside and cooperating in a European context, Switzerland sought a certain distance, trying not to be swept along by the big political currents, but maintaining its own traditions and identity on a small scale. With growing international interdependence, a vivid interest in the development of the world economy, and the close relationship of the Swiss economy, finance, and scientific research with their European counterparts, this reserved behavior and distance is coming under growing pressure. By now, "worldwide openness" as well as "narrow-minded isolation" are both Swiss trademarks.

Even if chances for a neutral state to stay outside armed conflicts during a future war in Europe are small, they are better than zero and better than those of the partners within the alliances. This is one of the main reasons why it is the Swiss people's firm conviction that today, and in the forseeable future, there are hardly any alternatives to a Swiss policy of neutrality worthy of serious consideration. In the act signed in Paris in 1815, Switzerland not only obtained the recognition of its "perpetual neutrality," but also the assurance of the European powers that this neutrality was "in the true interest of the policy of Europe as a whole." Is this still true in the sphere of today's security policy? Could Switzerland's contribution to the defense of Europe be greater or better than the one she is in a position to make by means of her national defense? Are the measures of Swiss security policy still a factor of stability in Europe, that is, a factor that counts and could therefore be understood as an act of "European solidarity?" In order to answer these questions, it is necessary to examine the security policy and military measures of Switzerland critically. Are the Swiss "hangers-on, profiteers, outsiders" in the field of security policy, thus violating the principle of European solidarity, or do they contribute significantly to the defense of European key positions? It is not by chance that the discussion of these points and the world political crises of past years have increased the weight attached to security policy in this country, and to its main element—armed neutrality.

The Security Policy of Switzerland

We live in a time of contrasts concerning the general politico-military situation. On one hand, many states are augmenting their economic, scientific, and technical collaboration. The integration of markets is progressing steadily. On the other hand, strong ideological and social tensions—as well as those due to power politics—still exist. Increasingly, conflicts between states are not only carried out militarily, but also by psychological, economic, and terrorist means, or by indirect warfare.

The security policy of Switzerland takes into consideration both tendencies. By tradition, Switzerland welcomes all serious peace efforts. However, Switzerland would jeopardize its existence if it did not recognize that the threats mentioned above can also directly or indirectly affect a small neutral country. Certainly, the Swiss would not describe Switzerland as did the Austrian Hans Thalberg recently. "Switzerland's permanent neutrality is remarkable not only for its historic roots, going back over four centuries, but also for its unique international setting. Situated exclusively among like-minded countries of Western orientation, she is far removed from East-West rivalries and has little to fear for herself."[3] Quite to the contrary, the Swiss are convinced that efforts to defend themselves against force are a necessity. These efforts must be integrated within a comprehensive security policy having clear guidelines.

Based on the Swiss constitution, four security policy objectives are in effect. These fall under four topics.

Preservation of peace in independence. The preservation of peace is not an end in itself. It can neither be separated from the preservation of Swiss self-determination (meaning the freedom to order one's own affairs), nor can one be played off against the other. Switzerland's goal is "peace in independence." Both aspects are therefore of equal importance and have a price: Switzerland must be vigilant and willing to defend them. This does not exclude the exploration of additional ways and methods to ensure the preservation of peace.

The preservation of freedom of action. Independence presupposes the freedom of action to take those domestic and foreign policy measures corresponding best to Switzerland's political will and security needs. Freedom of action presupposes the availability of powerful means to withstand foreign pressures. By having those means at its disposal, Switzerland also complies with international law, which compels the permanently neutral state to reject demands and impositions of any kind and origin, including those cases in which they are accompanied by threats or the employment of force. The authorities must also have freedom of action in the interior of the country. If attempts are made, through illegal activities or even by force, to chance the democratic order based upon the will of the majority, the security of the population and of the state is put into question. To the extent that its main mission allows, the army helps the civilian authorities in case of massive attacks against the public order, if the available police forces are not able to repel them.

Protection of the population. The third security policy objective is to protect the population from the direct and indirect effects of weapons, conventional or nuclear, and from the effects of a military occupation of the country.

Defense of the territory. In case of an attack, Switzerland must maintain its territorial integrity to the fullest possible extent. The airspace above will be defended by keeping it under Swiss control. All four security policy objectives are unconditionally based upon the prerequisite that the army is capable of implementing the strategy.

The Strategic Objectives of Switzerland

The strategic posture of Switzerland is the result of the comparison of its security goals to the threats to which they are exposed. A realistic assessment of the situation shows that such threats to Switzerland are possible, and that some of them already exist.

The security policy distinguishes between four forms of military threat: (1) indirect warfare, such as psychological warfare, subversion, sabotage, and terrorism; (2) the threat on the conventional level; (3) an attack with weapons of mass destruction, such as nuclear, biological, or chemical weapons; and (4) blackmail, a possibility that exists at each level of conflict. These threats arise not only as a result of the unknown intentions of potential opponents, but also as a result of the basic East-West political-ideological tensions, and as a result of the existence of troops and armaments beyond purely defensive needs. These armed forces could be employed rapidly, and perhaps without or with very little or no warning, to achieve strategic surprise. Opinions differ concerning the probability of such developments, but not with regard to the possibility of the threat.

The determining factor for Swiss preparations must be the existing levels of foreign forces which are presently characterized by a high readiness of strategic weapons and of armies trained and equipped for offensive actions in Europe.

In most of the possible conflict scenarios Switzerland would likely be of secondary or tertiary interest to the aggressor. It is therefore neither right nor sensible to compare the full military might of a potential opponent to the defense forces of Switzerland. Nevertheless, Switzerland could become involved in a conflict, or be affected by the repercussions of one in which it is not the main objective. If defense preparations are to be timely and sufficient, a small country must take "worst case scenarios" into consideration in order to answer the question, "How much defense is enough?"

Although there are significant differences between the geostrategic situations of Switzerland and Austria, they can be classified together functionally. From a NATO viewpoint Austria and Switzerland form a "transverse barrier" separating NATO-center from NATO-south, driving a wedge between its defense line and thus opening a potential corridor of more than 800 kilometers in length. Important north-south lines of communication traverse Austria and Switzerland. Thus, a well-defended Swiss territory responds to a NATO concern that hostile forces would use the neutral country to conduct a bypassing maneuver in order to advance rapidly into the flanks or the rear of the forward defense positions.

From a Warsaw Pact viewpoint a strongly defended neutral Austria and Switzerland offer the advantage of a secure southern flank for an attack into the central region. After a successful campaign against the NATO forces Austria and Switzerland could be dealt with later. As Suvorov has written: "It must be emphasized that the task of the C-in-C of the Western Strategic Direction is to advance swiftly westwards and to concentrate all his efforts on this and this alone. He is covered on the south by neutral Austria and Switzerland, which, it is planned, will be liberated somewhat later."[4] At the same time, in the context of a surprise attack followed by a rapid westward thrust of Warsaw Pact forces, it would be tempting to use the Austria-Swiss air corridor in a first phase of a conflict to project hostile air power into central Europe and behind the main defense belt of NATO.

For NATO and the Warsaw Pact this corridor could therefore be of strategic importance on a European level, either for a coherent NATO defense line or for coordinated offensive operations by the Warsaw Pact. In every scenario the time element is of crucial importance. With strategic surprise the correlation of forces can be effectively upset in order to win a war in its opening phase. This is particularly true if a surprise attack is launched against a militia army which has few standing forces and mobilizes relatively slowly.

The objective of a land attack on Switzerland could be to force one's way through Swiss territory to open passageways for further operations. In this case the attack is likely to take place in only one direction, namely through the lowlands, or the "Plateau." If the opponent's goal was to conquer Switzerland at a later stage of the war, then the Swiss might have to deal with concentric attacks, launched from different directions. The opponent might attempt to penetrate rapidly the Border Brigades and to defeat the main forces on the Plateau in order to reach its operational goals.

A glance at central European topography reveals that a military operation crossing Switzerland would probably choose the Plateau as an axis of advance. The Plateau presents itself as a funnel which becomes narrower in the east-west direction, opening a gap between the Alps and the Jura Mountains, including the Swabian-Franconian Jura. A map also shows that the Plateau is primarily suitable for a thrust from east to west because the widest opening of the funnel can serve as an assembly area for the attack echelons. In addition, such an east-west thrust could serve the important operational goal of bypassing the Black Forest and the Vosges Mountains on the southern side in order to reach the Rhône Valley and Lyon. Such an advance has to be seen in the context of possible airborne operations using existing airports on the Plateau such as Kloten-Dübendorf, Emmen, Belpmoos, Payerne, and Geneva. The blocking of the lowlands, and especially the protection of landing zones against airborne operations, is a task with which Switzerland could be confronted overnight. The assessment of the central European topography makes evident that the defense of the Swiss Plateau is not solely in the Swiss interest. The effectiveness of the defense of the Plateau is of utmost strategic importance for Switzerland's European neighbors, especially France and Germany.

As far as defense policy and the conduct of war is concerned, the Swiss army master plan states clearly that a strategic offensive would most likely emanate from a totalitarian power. It has to be expected, therefore, that a conventional attack would be preceded, accompanied, and also followed by indirect warfare. The importance and danger of sabotage, terrorism, and special operations by commandos are continuing to rise, and the anticipated psychological effects increase accordingly. It is very likely that superpowers would try, in a state of increased tension or serious disruption, as well as in an open conflict in Europe, to misuse the obligations and duties of neutrals for direct political pressure. For example, the military incapacity of a neutral to detect and destroy low-flying cruise missiles could be used as a basis for massive threats.

Even though a direct attack against Switzerland with weapons of mass destruction is unlikely, a war with biological, chemical, or nuclear weapons cannot be excluded as long as any power possesses such weaponry. The spectrum of possible employment of such weapons has been steadily widened; technical improvements make feasible the pinpointing of targets and allow the limitation of unwanted secondary effects. Parallel to this development, the likelihood of using such weapons for political pressure has risen. This would be particularly effective against neutral states that lack means of retaliation. If these weapons were employed in an attack against Switzerland, primary targets might be communications installations, air force bases, SAM batteries, artillery concentrations, antiaircraft and mechanized reserves, as well as important logistical installations. The use of chemical weapons also would be expected in a so-called conventional war since they are well suited to the attack of selected targets with maximum surprise and a minimum of unwanted destruction. Where weapons of mass destruction dictate the rules of waging war, troops have to be trained and prepared to survive the effects of nuclear, chemical, and biological weapons, and to retain their fighting capacity. Even more important is the consideration that Switzerland must prevent, by all means, the use of nuclear weapons on its soil by countries trying to block an attack across Switzerland. A strong Swiss defense posture is the best insurance against such contingency nuclear planning by third parties.

Switzerland also has an unavoidable role in the "politico-military target area" of an East-West conflict. Essentially, the entire concept of neutrality relative to ideology is foreign to Marxism-Leninism. In the dialectical process of the ongoing class struggle all human beings are bound to give support to one side or the other. One cannot make oneself neutral by deciding to opt out of the class war altogether. Not to support the proletariat in his fight against the bourgeoisie is to help the latter. "He who is not with me is against me."[5] Switzerland, as an exponent of capitalism, is well situated in the "politico-military target area" of an East-West conflict. The determined defense of neutral Switzerland against any incursions from the East would be considered not a neutral obligation, but as open support for the opponents of the Warsaw Pact. This "objective solidarity" with NATO states exemplifies an ideological unity from the Soviet view. Switzerland is a firm component of the capitalist camp. As recently written: "The Russians obviously regard arms neutrality as a charming paper phrase. They associate Switzerland with the North Atlantic Treaty Organization, even if Switzerland protestingly does not. The Russians appear to look upon Switzerland as a kind of capitalist Alamo—a likely position of ultimate defense for a falling Western Europe."[6]

One of the guiding principles of Swiss security policy is that if Switzerland is involved in a war, then it is no longer bound to its neutrality and to those foreign policy and military limitations associated with this status. In such a case Switzerland would be free to collaborate militarily with the opponent of the attacker. This principle might be interpreted by the Soviets to mean that Switzerland will be treated the same as NATO states from the moment that its neutrality has lost its utility for the Soviet Union. The only difference that counts is the fact that Switzerland has no nuclear weapons. In effect, the usefulness of the neutral status of Switzerland might be of limited duration. If ever the political decision is made to occupy Europe or parts of Europe by military means, then the ideology embedded in the Marxist-Leninist views of Soviet military doctrine dictates that the neutral states will be treated as equal to the NATO states. The respect for neutral status will be subordinated to the politico-strategic goals. These will determine, along with the defense preparedness of the respective neutral country, whether it is included in the strategic/operational planning.

In the Swiss concept of general defense six main strategic missions are designated. Due to the possible overlapping of different types of threats, Switzerland must be in a position to fulfill these missions simultaneously under certain circumstances.

1. General defense and the preservation of independence in a state of relative peace
2. The securing of peace and crisis management
3. War prevention through defense readiness (dissuasion)
4. Conduct of war
5. Damage limitation and the guarantee of survival
6. Resistance in occupied areas

The main emphasis of Swiss security policy and of the strategic measures is upon the third mission—to prevent the involvement of the country in a war through defense readiness. Some strategic means are seen as more appropriate than others for dealing with certain strategic missions. For example, the strengthening of the armed forces increases not only their chances of victory in a military conflict, but also their "dissuasive effect," making it more likely that the country stays out of war.

Dissuasion (or inoffensive deterrence) is a strategic posture that should persuade a potential opponent not to initiate an armed conflict against Switzerland. The opponent should be convinced of the disproportion between the advantages gained from an attack and the risks entailed. I described this earlier as "keeping the price of entry as high as possible." The risks which a potential aggressor must face are the loss of military forces, which would affect the overall correlation of forces, prestige, and time. A war can be prevented through deterrence, political measures, and defense readiness. Deterrence as a strategic posture (consisting of a credible threat to retaliate) is beyond Switzerland's capacity. A political measure might be the attempt to influence a potential aggressor by showing him the advantages of an intact Switzerland. In the case of an open conflict involving third powers it would be the task of Swiss foreign policy to bring about the recognition of Switzerland's neutrality by all belligerents. Defense readiness consists of the visible determination and an appropriate armament to withstand an enemy attack even if the most powerful means are employed, not only against the armed forces, but also against the population. This readiness must have the highest priority.

Switzerland's goal is to continuously improve and strengthen its defense measures in order to keep out of war. The term "dissuasion" applies to more than just military efforts. A dissuasive effect can be attained only on the basis of combined efforts in the military and civilian spheres (concept of general defense). Dissuasion depends upon a far-reaching capacity for endurance on the part of the population and the authorities. Naturally the assets of a small country do not weigh heavily in a European or even world theater of war, but they can influence the aggressor's assessment of the situation in such a way as to dissuade him from attacking the country.

If the strategy of dissuasion fails, and Switzerland becomes involved in an armed conflict in Europe, or is directly attacked, the goal is to ensure the survival of the population and the nation through defensive military measures. There will be no question of symbolic resistance, but of a strong, forceful, and determined defense. Swiss defense measures are aimed at yielding the least amount of territory. If an opponent succeeds in occupying parts of Switzerland, an opponent can expect not only passive but also active resistance. Preparations for this resistance have been established in peacetime.

The army is the most powerful element at Switzerland's disposal for the implementation of this strategy. The Swiss depend upon it to effectively oppose an armed attack. The major characteristics of the army are compulsory military service and the militia system. These insure a quantitatively strong army. Thus, a main asset of the army is the large number of soldiers who can be mobilized. The fact that so many people (10.7 percent of the population) are involved in the army creates a bond between it and the population, making the army a factor of national cohesion. The army is essentially structured as a fighting instrument. It uses practically all of its training time in preparing combat missions within its own territory. Additional obligations are minimal.

The army accomplishes its strategic mission of preventing a war by (1) making it clear to any potential opponent that in the case of an attack against Switzerland, the opponent must expect high losses of equipment and men as well as the destruction of all those military and civilian installations which could be of interest to him; (2) making it clear that a potential aggressor cannot achieve a fait accompli with a surprise attack, owing to prepared defense of key areas; and (3) dampening the hopes of any potential opponent for reaching his goal—whether it be occupation of territory, passage of troops, or use of airspace as a corridor—within a short time or at an acceptable cost. Should Swiss operational formations cease to exist as coherent, effective fighting forces, the army will transit to guerrilla warfare in order to prevent the opponent from gaining control over the occupied territory, and to prepare for its liberation.

The terrain in Switzerland has an important influence on the operational art and on tactics. The most important features of Swiss territory, as related to operational zones, can be divided into three main regions: the Jura, situated in the northwestern part of the country, cover about 10 percent of the surface and have peaks up to 1,600 meters; the Alps cover the southern part of the country, about 60 percent of the land, and have peaks up to 4,600 meters; and the lowlands (the Plateau) cover about 30 percent of the land and contain rolling hills between 400 and 600 meters in altitude. These three areas become operational zones as the border zone, the lowlands, and the central zone.

The Jura, or border zone, consists of oblong hills and covers the northern and northwestern border. Any aggressor attacking through this area cannot advance in a wide front to the center of Switzerland. He will have to force his way through numerous bottlenecks and defiles which are easily blocked. A considerable number of fortifications as well as artificial obstacles have been set up to hamper his progress. Compared with the Jura, the Alps form a much stronger and more natural obstacle. Mechanized operations can be carried out only in some of the wider valleys. Most of the alpine zone requires the commitment of strong infantry forces. Large vital parts are conquerable only by specially trained and equipped mountain troops and/or possibly by air mobile forces. Furthermore, weather conditions (such as frequent snow) change the topography and hinder operations. One must reckon with such a hindrance in the Jura each year for four months, and for eight months in the

Alps. Therefore it can be assumed that the main thrust would be directed along the valleys in the central part, this being the location of the road and railway net. Consequently, the doors to the Alps and the center are reinforced by fortifications. The only region offering suitable conditions for mechanized operations are the lowlands, or Plateau. However, even this area cannot be called open country. In the east and center it is about 45 km wide, in the west about 35 km wide, and it is crossed by numerous rivers. At about every 10 km, a river runs at right angles to the most likely axis of advance. But because most of the cities are part of the lowlands, this part of the country is of particular importance.

The "high price of entry" is determined not only by the terrain, but also by the permanent infrastructure. Much of it is prepared with demolitions. There are more than two thousand permanent demolition devices installed in important bridges and objects, four thousand permanent obstacles, tank barriers, and other blocks to hinder mechanized movements. In two thousand fortified areas there are more than three thousand weapons systems. Two thousand shelters for 20 percent of the troops are ready. Command posts and other installations for command and control are built in concrete dugouts or under rock.

With the well-developed permanent infrastructure, a thoroughly prepared system of mobilization, and skillful use of the advantages of the terrain, the armed forces can reach combat readiness quickly. The mobilization system allows the government, depending upon the threat, to call on limited duty parts or the whole of the armed forces. The system can mobilize more than 676,000 men within two to four days—that is, about 11 percent of the Swiss population, or about 28 percent of the male population.

CONCLUSIONS

I have tried to demonstrate that the strong determination of Switzerland to defend itself—and at the same time, to contribute to the defense of key European positions—is still intact. If it comes to war, Switzerland is a nation at arms with a population mobilized for comprehensive defense. Switzerland, like Sweden, has air and armored forces that, relative to the size of the population, ranks as the largest of Western Europe. As far as Switzerland is concerned, there is no "soft underbelly of the Central Front." NATO may indeed be better served by Switzerland staying outside NATO than by entertaining the idea of including it in the alliance.

The questions set forth early in this chapter—as to whether the security policies and military effort of a small neutral country like Switzerland contribute to stability in Europe and can thus be understood as an act of European solidarity—remain open. The only real test will occur in a crisis—something which, fortunately, has yet to be faced.

CHAPTER FIVE

YUGOSLAVIA

Jens Reuter

The important geostrategic position of the Balkans makes East-West military confrontation and superpower competition in the region, as well as in the Mediterranean and the Middle East, relevant to the security of all Balkan countries. Yugoslavia's security has been, and will continue to be, determined by the nature and intensity of NATO-Warsaw Pact rivalry in the region.

The policies adopted by members of each alliance toward Yugoslavia are in turn a function of how each perceives the threat from the opposing alliance. Yugoslavia, after having been expelled from the Soviet-dominated "socialist camp" in 1948, benefited from the fact that the West perceived the move as a serious blow to the Soviet Union and hence deserved Western and particularly U.S. support. Indeed, Moscow lost its influence in a region of growing strategic significance. Defense arrangements within this sphere of southeastern Europe could no longer be extended via Yugoslavia to the geopolitically important Alps and the Adriatic Sea. A new defense line had to be established much farther behind, in Hungary and the Balkans, isolating at the same time the still Moscow-oriented Albania.

The Western alliance came to regard nonaligned Yugoslavia as a protective shield for Italy and a defensive barrier against a possible Soviet drive through the Balkans to the Mediterranean. Yugoslavia also provided the so far missing link to Greece and Turkey on NATO's southeast flank. Western plans to include Yugoslavia in NATO (mainly by way of a tripartite pact with these two countries), however, failed. In 1955 Krushchev went to Belgrade and offered reconciliation. Still, Yugoslavia's nonaligned status disadvantaged the Soviet Union more than the West because, as a transit or base area for military operations, the country is of greater importance to the Soviet Union than it is to NATO.

As long as Yugoslavia remains outside the Warsaw Pact she helps to protect northeast Italy from a surprise offensive by land, prevents the Soviet Union from direct access to naval bases in the Adriatic Sea, and reduces the threat from the Soviet air force. Yugoslavia, therefore, is playing an important role in Western military planning as a territory that constitutes a buffer zone between the Warsaw Pact forces in the Balkans and the NATO forces in Italy and Greece, the central and western Mediterranean.

The expansion of Soviet global military capabilities in the seventies further increased Yugoslavia's strategic significance for the Warsaw Pact. It was seen as a bridge or springboard to the Middle East and the entire Mediterranean. The country's strategic value for the Soviet Union was evident during the 1973 Arab-Israeli War when the Soviet Union was allowed to make use of Yugoslavia's airfield and the harbor of Rijeka in order to ship heavy military equipment. Soviet naval operations in the Mediterranean have always been constrained by a lack of naval bases and facilities. When the Soviet Union was forced to abandon Valona Bay in Albania (1961), her first naval base in the Mediterranean, and naval support facilities in Egypt were only of temporary use, she turned her attention to Yugoslavia. The Soviet navy made maximum use of Yugoslavia's repair yards, which are open to all countries. In 1974, however, Yugoslav legislation limited the rights of access to Yugoslav ports. Even more importantly, Yugoslavia's readiness to ad hoc cooperation in 1973 did not lead, as many Western observers feared, to closer military links between Belgrade and Moscow.

Soviet attempts to push rapprochement to the point of reincorporating Yugoslavia, even partly, into the Eastern military alliance are bound to fail as long as the Soviet Union is regarded by Yugoslavia as its only direct military threat. Yugoslavia's defense policy is deeply rooted in the experiences of World War II. It is feared that, not unlike the German invasion of 1941, the country could be invaded by conventional forces of one of the great powers. At this time the most likely aggressor is the Soviet Union; Yugoslav defense planning treats her in an almost analogous fashion to Germany and Italy during World War II. Although the danger of Soviet invasion seemed real only during the first three years after Yugoslavia's expulsion from the socialist camp, old fears die hard. The Soviet invasion of Hungary in 1956 and that of Czechoslovakia in 1968 were played up for a number of reasons, domestic and external, but were not seen as a direct threat to Yugoslavia.

In spite of the fact that Yugoslavia's defense policy originates in the country's experiences with guerrilla war against the occupying power during World War II, a large conventional standing army (People's Army, YPA) was set up. Its tactical doctrines were primarily concerned with the static defense of cities, and other fixed positions took charge of Yugoslavia's security. After the invasion of Czechoslovakia in 1968, a mainly politically motivated change in military and tactical doctrines was initiated. The new concept of "All-People's Defence and Social Self Protection" was laid down in the All National Defence Law (1974). It provided for basic strategic defense for the entire country with regular forces being used alongside a citizen's territorial army. The newly established Territorial Defence Forces (TDF), unlike the conventional YPA, are compared to, and tactically inspired by, the wartime Partisan Army. The structure of TDF is decentralized with civilian authorities at republican and communal levels playing an important role.

The idea of combining elements of both traditional military and partisan strategies is not undisputed. It caused considerable problems within YPA's leadership. The 1974 constitution confirmed the All-People's Defence but clarified the command structure in favor of the authority of the federal army. The intention of All-People's Defence so far is to increase the "price of entry" and to demonstrate Yugoslavia's readiness to defend her independence, as Pierre Maurer pointed out.[1] This intention finds its most determined expression in the Yugoslav Constitution of 1974 (Article 238), which expressly forbids military capitulation.

Recent events in Yugoslavia have highlighted a dimension of Yugoslav security that under the severe leadership of President Tito had little relevance, namely, the problem of internal cohesion and stability. Today, some years after Tito's death, Yugoslavia is faced with numerous latent or open tensions and conflicts: a serious economic crisis, unrest among the minorities, tensions among individual republics, and an ever growing criticism of the communist leadership. These developments would confirm a more general trend in quite a number of countries which, not unlike Yugoslavia, are confronted with serious economic problems and have either a federal constitution and/or various ethnic or religious minorities. The combination of economic failure and growing political restlessness makes an explosive mixture that, particularly in a period of détente, turns attention away from concerns about external security to those about internal stability. In such a situation the question of the role of the armed forces almost inevitably shifts its focus: their role is seen to be more one of assisting in maintaining internal order rather than external defense. These are obviously politically highly delicate issues. But they also have come to the fore in Yugoslavia. Any consideration of Yugoslavia's security policy today must therefore encompass this domestic dimension. It was not the object of the present contribution to do this in any detail. But it certainly deserves to be kept in mind when Yugoslavia and the future orientation of its security policy are being examined.

THE ORIGIN OF NONALIGNMENT

Yugoslavia's defense policy is deeply rooted in the experiences of World War II. The country had fallen victim to a great hegemonic power, but liberated itself in a long and cruel partisan war. Socialist Yugoslavia entered the international stage with the consciousness that a small country was able to fight successfully against an invading power. Her leadership developed a powerful feeling of pride and conviction that Yugoslavia did not owe anything to anyone, because she had given more than her share to the victory over Nazi Germany. Having behind him the strongest army in Eastern Europe, Tito claimed for his country the status of a victorious power and, more concretely, Trieste and territory in Carinthia (Austria). "We have come to annihilate and help destroy the greatest enemy of civilisation—Germany. We have a right to remain there [Trieste] as allies, because we want the recognition of our allied rights, just as we have acknowledged our own duties that we have fulfilled one hundred per cent and are still doing so."[2]

Tito's demands not only damaged Yugoslavia's relationship with her Western allies but it also countered Soviet plans for bringing Yugoslavia in line with the East European countries liberated and occupied by the Red Army. Yugoslavia's territorial claims regarding Italy and Austria, as well as her involvement in the Greek Civil War, did not suit Stalin's strategy. Hence, his support for them was only lukewarm. The discovery that the Soviet Union was more a threat than a protector came as a shock to the Yugoslav communists. Signs of incipient disagreement between Yugoslavia and the Soviet Union could, however, be detected already in various incidents during the war. But for reasons of ideological affinity, the Soviet Union was considered to be Yugoslavia's natural ally. It was no surprise, then, that Belgrade came out strongly against the Marshall Plan. The Yugoslav communists wanted to develop the country solely with the aid of the socialist bloc. Yugoslavia's economy was thus to be built along socialist lines on the assumption that large-scale trade with, and aid from, the Soviet Union would follow.

In the military field, too, Tito quickly turned his back on his wartime Western allies and looked for arms, equipment, and guidance from the Soviet Union. Since Yugoslavia had decided to convert partisan units into a regular army, she needed assistance. This had already become apparent in the final phases of the war, when Moscow offered valuable help in arms and supplies along with providing tactical cooperation with the Red Army. Military leaders and younger officers were sent for training in Soviet military academies and Soviet military advisers came to Yugoslavia. Cooperation was based on the Treaty of Friendship and Postwar Collaboration between the USSR and Yugoslavia, signed by Molotov and Tito on April 11, 1945.[3] Five days later, Milovan Djilas stated: "There is no force on earth which could break the fraternal alliance of the peoples of Yugoslavia with the people of the Soviet Union. . . . It must be remembered that indissoluble friendship with Great Russia has been the age-old dream of all southern Slavs."[4]

Such a close relationship between Yugoslavia and the Soviet Union, however, raised a number of questions in the field of intelligence and state security. The Yugoslav leadership had to learn that Stalin was interested less in partnership and more in subordination. The idea that the interests of socialist Yugoslavia must necessarily coincide with those of the homeland of socialism proved to be a fiction: what was good for the latter was by no means always good for the former. The Yugoslav communists came to make a clear distinction between their national interests and those of the Soviet Union. In April 1948 Tito wrote in a letter to Stalin and Molotov, "No matter how much each of us loves the USSR, the land of socialism, we in no case love our country less."[5]

Several works have already addressed the so-called Cominform conflict. What is important in the present context is that since the Cominform conflict, the USSR is perceived by the Yugoslavs as a great power like any other—in other words, a hegemonic force threatening the independence of small countries.

The Tito-Stalin controversy brought about a turning point in Yugoslav security policy. From then on, Yugoslavia took the Soviet Union to be the most serious, and later on, the only, threat to Yugoslav security; she must prevent the Soviet Union and her allies from attempting to carry out acts of aggression against Yugoslavia similar to those levied against Hungary in 1956 and Czechoslovakia in 1968.

But the Cominform conflict is also a source of Yugoslavia's conviction that there is no need for small states to submit to great powers. Dealing with Soviet pressure on Yugoslavia, Mose Pijade, the party's leading theoretician, wrote in July 1949: "There is no justification at all for the view that small nations must jump into the mouth of this or that shark. If that were a social law, there would not today be any small states."[6] The Yugoslav leadership, however, knew that self-confidence alone would not defend the country. Tito's famous words, that "we should rather go hungry and barefoot than sacrifice our independence,"[7] were nothing but words for the gallery. Until 1951 the danger of Soviet invasion remained serious. By withholding arms and supplies from Yugoslavia's war industry, Moscow tried to weaken the country's capacity for defense. Besides that, Yugoslavia suffered from the economic blockade imposed on her by the Soviet Union and its allies.

RAPPROCHEMENT WITH THE WEST

At the beginning of the Cominform conflict, and as a result of Yugoslavia's support for the Greek partisans and the unresolved conflict over Trieste, relations with the West were at a low ebb. But the dispute forced Yugoslavia to seek better relations with the Western states. Even though she persisted in fighting, at least verbally, "Anglo-American imperialism," she soon obtained aid from the West. She thus became one of the first beneficiaries of the cold war, with its fierce rivalry between East and West. The Western powers soon came to realize that the existence of an independent socialist Yugoslavia was not only in their own interest but had also a disruptive potential for Eastern Europe.[8] Even more important, it became clear that Yugoslavia might serve as a barrier to any Eastern military advance toward the Mediterranean. The Korean War and Cominform propaganda sharpened the Western military planners' and politicians' sensitivity to the threat to Yugoslavia from the Soviet Union. On 29 November 1950 President Truman stated: "The continued independence of Yugoslavia is of great importance to the security of the United States. We can preserve the independence of a nation which is defying the savage threats of Soviet imperialists, and keeping Soviet power out of one of Europe's most strategic areas. This is clearly in our national interest."[9]

There is no doubt that Stalin, who expected Tito's surrender or his deposal by the Moscovite faction of the Communist party of Yugoslavia, would have been winner of the match had it not been for Western economic and military aid. Yugoslavia could defend her independence against the Soviet Union only by cooperating with the West. This, however, reached a stage at which it threatened Yugoslavia's independent status.

For a time it seemed that Western military aid might bring Yugoslavia into NATO in at least an associated status. In February

of 1953 Yugoslavia concluded the Treaty of Friendship and Cooperation with Greece and Turkey. This treaty provided the basis for
informal consultations among the three general staffs of the three
countries.[10] This first treaty with Greece and Turkey grew into the
Treaty of Political Cooperation and Mutual Assistance in August
1954.[11] The ultimate idea of the NATO strategists was to link Yugoslavia to NATO countries so as to create a defensive barrier against a
possible Soviet drive through the Balkans to the Mediterranean.
This was more than Tito wanted to concede. Yugoslavia should
not be an instrument of any superpower's politics. First steps in
Yugoslav-Soviet rapprochement after Stalin's death, and the deep-
ening quarrel between Greece and Turkey over Cyprus, made it
possible to quit the tripartite alliance—which was still listed in a
reference book register in 1968 as a treaty with NATO.[12]

Though nonalignment had not yet crystallized into a new
political doctrine, nonparticipation in any military alliances dominated by great powers had become a basic principle in Yugoslav
security policy. The Eastern bloc was viewed as a substantial threat
to Yugoslavia's security. The Western alliance, in spite of protecting
the country's security, was also considered a threat to Yugoslavia's
independence, in as much as it seemed to narrow her political room
to maneuver. To accept Western economic and military aid did not
mean that communist Yugoslavia would become part of the West.
After having resisted Soviet pressure thanks to Western help, Yugoslavia made it a basic goal of foreign policy to defend the country's
independence against the West. The first public confession to a
policy outside both blocs was presented in a speech, given by Edvard
Kardelj at the fourth session of the UN General Assembly in 1949.
"Yugoslavia does not belong to any military bloc nor will she be a
participant in any kind of aggressive planning against any country."[13]
One year later, at the UN General Assembly, Kardelj commented
on Yugoslavia's challenge to both blocs.

> The peoples of Yugoslavia cannot accept the assumption that
> mankind must today choose between domination by one or
> another great power. We consider that there is another
> road, though difficult, is nonetheless the necessary road of

democratic struggle for a world of free and equal nations, against interference from outside in the internal affairs of nations and for an all-round peaceful cooperation of peoples on the basis of equality.[14]

THE RISE OF NONALIGNMENT

The early fifties were, in Rubinstein's words, the period "between unalignment and nonalignment."[15] Yugoslavia was looking for a theoretical justification for her political course between East and West. In short, she sought a strategy to find new allies. A major step was the discovery of the United Nations as a tribunal for accusing the Soviet Union and, consequently, for gaining widespread support. Rubinstein noted that at this time any additional margin of security, however tenuous, was desperately welcome.[16]

Yugoslavia's election to one of the six nonpermanent seats on the Security Council in October 1949 helped to develop her new relationships. The UN also served as a forum for arousing international public opinion. In the same month Yugoslavia tabled a Declaration on Duties and Rights of States which already included the principles that later became part of Yugoslavia's policy of active and peaceful coexistence.[17]

Yugoslavia supported the United for Peace resolution in October 1950 after the outbreak of the Korean War. She then proposed a resolution of her own on the Duties of States in Event of the Outbreak of Hostilities. It was based on the declaration of 1949 and was passed by the General Assembly on 17 November 1950.[18] In November 1951 Yugoslavia lodged a formal complaint with the United Nations against the hostile actions of the Soviet Union.[19] Helping to pioneer small nations' use of the UN as a restraint on the actions of great powers, Yugoslavia built bridges to the Third World through UN meetings.[20]

As Yugoslavia became more and more aware of the importance of the newly independent nations, a new strand emerged in her foreign policy. First of all, Yugoslavia came to see that the young Afro-Asian states were potential allies in her struggle for "un-engagement." She began to perceive international developments not only in terms of East-West conflicts, but also in terms of north-south problems. Her "fight against imperialism and colonialism" motivated by ideology and combined with her "fight for disarmament and development" opened the door to the Third World. After one Third World trip, Tito spoke at a meeting in Belgrade about an alternative to alignment with one of the hostile blocs.

> The world is today divided in two blocs, but it is fortunate that in addition to those two blocs there is an enormous number of men and states who believe that it is a wrong policy to follow the line of division and not to do anything for the line of integration. . . . Ideological differences cannot be the reason for division of the world into blocs which would whet their knives.[21]

The alternative was seen in a policy rejecting any membership with a military bloc and in "associating with other countries having similar views in an active effort to improve international relations."[22] In India, Tito had declared that his country wanted to increase the number of states and nations "who place the safeguard of peace above all else and who struggle for the active co-existence of states with different social systems."[23] From this moment onward, "peaceful and active co-existence" became the essential ingredient of Yugoslavia's policy of nonalignment.

The communiqué of Tito's meeting with Nehru stated that the new policy was not synonymous with "neutrality" or "neutralism." For these two leaders, this was not a passive or evasive policy, but rather it was "an active, positive and constructing policy, seeking to lead to collective peace on which alone collective security can really rest."[24]

According to Leo Mates, this new policy is defined negatively "as a policy of non-participation in bloc groupings, military alliances, or political blocs."[25] In 1961, at the preparatory meeting for the first nonaligned summit, five criteria of a nonaligned policy were formulated: (1) to follow an independent policy based upon coexistence and nonalignment; (2) to support national liberation movements; (3) not to become a member of any military alliance; (4) not to enter into a bilateral alliance with a superpower; and (5) not to accept military bases of a superpower on one's territory.[26] A sixth criterion, made in Yugoslavia, should be added: not to serve as a member of an economic grouping that is headed by great powers which are closely related to or integrated with military alliances.

One final point which is an inherent part of Yugoslavia's policy of nonalignment is that cooperation among the nonaligned countries can be used to strengthen their own international status and to improve their security. Yugoslavia strongly promoted the development of the movement of nonaligned countries. In Belgrade in 1961 she hosted the first Meeting of the Heads of State or Government of Non-aligned Countries. The movement was seen as a support of her policy outside of two military blocs. The growth in the number of independent states gave Yugoslavia the chance to develop this simple national goal into a general strategy of ever more nonaligned countries and a means to promote international peace.

Yugoslavia's leading role in the Third World helped to strengthen her position between the blocs. Using her standing with the nonaligned countries as a defense against pressure from East and West, Yugoslavia sought to increase her bargaining power vis-à-vis the two blocs. In this sense, Yugoslavia's approach to nonalignment served as a way to obtain security through gaining friends and prestige, in the face of lack of wealth.[27] Nonalignment and activity in the nonaligned movement can be seen as a strategy to protect national independence and territorial integrity by political means. In other words, nonalignment became a strategy to protect and justify Yugoslavia's unalignment.

NONALIGNMENT IN ACTION

There is no doubt that Yugoslavia's association with other non-aligned states increased her international prestige. But that did not solve the problem of Yugoslavia's need for economic aid. It was after World War II, when Stalin claimed subordination as a prerequisite for Soviet aid, that the Yugoslav leadership found out that a country in a state of economic dependence cannot defend its political independence. As a recipient of Western aid, Yugoslavia could avoid closer ties to the Western alliance only through better tactics. Cooperation with the West helped to uphold Yugoslavia's stance against the Soviet Union. But at the same time it risked undermining the country's unalignment.

After Stalin's death, therefore, Yugoslavia normalized relations with the Soviet Union. The rapprochement in 1955–56 made it possible to move to a middle position, with no binding tie to either side. She therefore played down the importance of military contacts with the West and pretended not to be dependent on Western military aid. This middle road has been a hard line to follow. The West first had to be convinced that good, even friendly relations with the Soviet Union would not bring Yugoslavia back into the Eastern orbit. As long as Tito was alive, his views on international affairs often brought him into conflict with the West. But on the whole Yugoslav-Western cooperation was free of severe conflicts, with notable exceptions. Problems emerged in the following situations: in connection with the Soviet invasion of Hungary; in 1957 when Tito recognized the GDR, knowing full well that this would deeply antagonize the Federal Republic of Germany; again, in 1961, when he praised the resumption of Soviet nuclear tests; and in 1967 and 1973 when he permitted Soviet planes to overfly Yugoslavia during the Middle East wars. His basic position was to speak openly against what he judged to be gross examples of "American imperialism." In the West, however, the interest in Yugoslavia's standing outside of the socialist bloc weighed heavier than the outrage over Tito's flirtations with Moscow or his schoolmasterly behavior. Tito learned to play American aid against Soviet hostility, and vice versa. To him the most dangerous times occurred during phases of Soviet-American rapprochement. Nevertheless, with the possible exception of the Carter administration, every American president, Tito himself, and his successors were and are aware of the common Western-Yugoslav goal: to prevent Soviet expansion into the Mediterranean.

The maintenance of good relations with the Soviet Union was a more difficult task. The events in Hungary in 1956 and Czechoslovakia in 1968 caused severe conflicts and reminded Belgrade that the Soviet Union was a serious threat. In both cases, the Yugoslav communists were forced into making concessions which damaged their credibility. Yet they did succeed in avoiding any long-lasting collisions.[28] Soviet assaults and Yugoslavia's fear that she might be the next victim of "fraternal help" revived Western interest in and understanding for her policy of nonalignment.

Both superpowers have been induced to consider Yugoslavia's nonalignment as a constituent part of the status quo in Europe, guaranteed by the United States, respected by the USSR, and accepted by the rest of Europe. But the country's deteriorating economic situation could jeopardize this balanced position. The problem is illustrated by catchwords like "foreign debt" and "Western credits" on one side, and "export dependence" and "energy dependence" on the other.

More seriously, Yugoslavia was not able to bring about balanced relations in the military field by means of diversifying the sources of her military equipment. While producing two-thirds of her requirements domestically, Yugoslavia remains dependent on the Soviet Union for most of her major arms, especially fighter bombers and SAMs. This reliance on Soviet military equipment began in 1957, when U.S. military assistance ended. Restrictions and delays in delivery were typical Soviet tactics to put pressure on or at least annoy Belgrade. Yugoslavia's development of military links to China, beginning in 1978, just like the cooperation with Romania, turned out to be relatively nonconsequential.

The acquisition of U.S. military equipment, first begun by Tito in 1974, is a delicate problem for both sides. Yugoslavia has, for the most part, no money for buying U.S. weapons and is not willing to increase her foreign debt by doing so. The Soviet Union, on the other hand, supplies arms and equipment as part of her bilateral trade. This means that Yugoslavia does not have to pay in hard currency, but simply with domestic products. In 1988 military imports from the Soviet Union even helped to solve what could have become for the Soviets an embarrassing situation. As a result of her declining income from oil exports—due to a fall in the price of oil— the Soviet Union incurred a huge trade deficit with Yugoslavia ($1,242 million).[29] The answer, of course, was to ship additional arms. Thus, in 1988, the Yugoslav air force proudly presented the latest addition to its arsenal: sixteen Soviet-built MiG-29 supersonic fighter bombers. For the time being, MiG-21s are the backbone of the Yugoslav air defense system, but they will be phased out in the 1990s.[30] Never before has Yugoslavia received Soviet equipment of such modern vintage and at relatively low prices. It is thus ironic that Yugoslav security is backed up by NATO against a potential Soviet threat, while at the same time it depends on Soviet equipment.

LINKING WITH THE INTERNATIONAL ECONOMY

Economic cooperation among nonaligned countries, as a strategy to ensure independence from both blocs, has been part of Yugoslavia's policy of nonalignment since the earliest days. Such an approach, however, has been the reality of its economic life. In spite of these policies motivated more by ideology than by economics, Yugoslavia has been forced by her weakness to seek privileged access to the major economic groupings: EEC and COMECON. They have become the most important of Yugoslavia's partners. The first serves as a creditor and supplier of capital goods and know-how, while the latter serves as an importer of products, which no country outside COMECON would pay for, as well as an energy supplier. It was thus only logical that Yugoslavia join not only UNCTAD and the Group of 77, but also OECD and GATT and develop special forms of cooperation with the EEC and COMECON.[31]

Initially, Yugoslavia was proud not to belong to any economic grouping headed by a superpower. Now she fears becoming more and more isolated in Europe. Unlike the European neutrals of Austria, Switzerland, Sweden, and Finland (associate member), Yugoslavia did not join EFTA. In 1992, when the European internal market is expected to be established, Yugoslavia (along with Albania) will be the only country in Europe which stands outside any economic grouping. In Brussels no less than in Belgrade the responsible politicians know that the future development of the Yugoslav economy depends on the EEC's readiness to give Yugoslavia some kind of preferential treatment and economic support. For now it looks as if the ties between Yugoslavia and the EEC are solid enough to assure the continuation of wide-ranging cooperation.

Fear of possible isolation has already provoked a lively debate in Yugoslavia as to whether she should apply for membership in the European Community. Such a step would mean the abandonment of a basic principle of Yugoslavia's nonalignment: nonaffiliation with economic groupings. The broad discussion in all mass media about the advantages and disadvantages of membership in the EEC clearly demonstrates that some forces in the Communist party are prepared to water down the concept of nonalignment. So far, however, a majority of the party still stands for maintenance of the status quo.

There can be no doubt that Yugoslavia's economy is in serious trouble. Burdened with $21 billion of foreign debt, she finds it increasingly difficult to defend her independence, both economically and politically. Urged by the Western creditors, the government initiated economic austerity programs one after the other. The International Monetary Fund is now playing a key role in Yugoslav economic policy. Backed by the commercial creditors, the World Bank, and Western governments, the IMF is in a strong position to demand substantial reforms in the economy. No wonder, therefore, that the federal government wants to "liberalize" the economy, with free prices for goods and labor, and private investment through bonds and shares.

EUROPEAN SECURITY

The idea of membership in the EEC, though rejected officially, indicates a new orientation in Yugoslavia's foreign policy—one pointing in the direction of Western Europe. The motivation for this change goes beyond economics; it has to do with the decline of the nonaligned movement and Yugoslavia's decreasing influence therein. There is little unity left as far as the basic orientation of the movement is concerned. Some years ago a minority of nonaligned countries headed by Cuba embarked on an attempt to maneuver the entire movement into closer ties with the Soviet Union. Yugoslavia realized that such a development would render meaningless the whole idea of the movement. Although Yugoslavia will be host to the movement's next conference, she has few illusions left as to its limited importance in world affairs.

The growing number of disputes among the nonaligned countries and the identity crisis of the movement were only part of the reason for Yugoslavia to look for new fields of political activity. Negotiations and meetings in the preparatory stage of the Conference on Security and Cooperation in Europe (CSCE) brought Yugoslavia to realize that, as a European country, her security problems could not be solved in the Third World. Increasing interest in European affairs has become the most conspicuous characteristic of Yugoslavia's post-Tito foreign policy.

Fearing that the CSCE might become a kind of "European concert"[32] led by the two superpowers, freezing the status quo and thus curtailing the room to maneuver of smaller countries in Europe, Yugoslavia tried to play an active role in the CSCE from the very beginning. She became a member of the N+N group, the coalition of neutral states (Switzerland, Austria, Sweden, and Finland), the nonaligned countries (Yugoslavia, Malta, and Cyprus), and micro-states like Liechtenstein and San Marino. Together with the other members of the N+N group, Yugoslavia helped to break deadlocks and to prevent a breakup of the Madrid review conference in 1982. In 1977–78, while hosting the first follow-up conference in Belgrade, Yugoslavia could not promote the process of détente in Europe as she had promoted nonalignment in the Third World sixteen years earlier. She came to learn that in the shadow of the superpowers, the possibilities and influence of a small state are limited. Still, Yugoslavia remained convinced that small states can have a tension-reducing function and thus she has become a major advocate of extending confidence- and security-building measures (CSBM) in Europe. In Madrid, Yugoslavia supported the French proposal for a conference on disarmament in Europe and put forward her own proposals for CSBMs. She urged stricter limits to the Final Act's provisions requiring advance notification of maneuvers involving more than 25,000 men. She has even pleaded for the limitation of military activities in general. Besides Malta, Yugoslavia played an important role at two meetings dealing with Cooperation in the Mediterranean Area (La Valetta, 1979, and Venice, 1984). Linking proposals for nuclear-free zones in both the Balkans and the Mediterranean, Yugoslavia pressed for the extension of CSBMs to those areas. The proposals met with opposition from the United States and indifference from the USSR. Both superpowers do not want to combine questions of European security with discussions of Mediterranean matters since, almost inevitably, delicate issues such as the Cyprus problem and the Near East conflict are linked to them.

Not being sure how cooperation in the N+N group would develop, Yugoslavia looked for new partners and for improving relations with them. First of all, despite severe differences over the Albanian minority living in Kosovo, Yugoslavia normalized relations with Albania. Thanks to skillful diplomacy, Yugoslavia hosted, in February 1988, the first meeting of Balkan states in which all countries participated. But Yugoslavia wants to go beyond that. She intends to widen this regional cooperation to security matters. Even if this should fail, the chances for widening cooperation in the nonmilitary fields are not bad. This in itself is important because Yugoslavia, with her eyes set on global and European objectives, has allowed relations with her immediate neighbors to deteriorate. As so often, she has had to recognize that security and stability begin at home and at one's door rather than in the lofty realm of global politics.

PART TWO

THE NEUTRALS AND EAST-WEST RELATIONS

CHAPTER SIX

NATO STRATEGY AND THE NEUTRALS

Lothar Ruehl

THE GEOPOLITICAL-STRATEGIC SITUATIONS
OF NEUTRAL COUNTRIES BETWEEN
EAST AND WEST IN EUROPE

Neutrality can have different effects on the political-strategic environment. It can result in a political balance between adversaries in a military conflict or, on the contrary, can tip the scales in favor of one side against the other. Neutrality changes with the evolution of conflicts and of the political-strategic environment. There is a considerable difference, in terms of security and foreign policy, between armed and unarmed neutrality. The reality and significance for other nations of a country's neutrality is largely determined by the overall distribution of military power as well as by the existence and nature of military or political alliances between neighboring countries. The geopolitical and strategic structure of postwar Europe determines to a large extent the specific situations and options of neutral countries. This is reflected by the political attitudes and the national defense postures of the neutral countries in Europe.

The political quality of neutrality is not defined by objective criteria; the term itself is ambiguous. The neutral countries of Europe call themselves, as a group in the CSCE and CDE negotiations, "neutral and nonaligned." Therefore, in their own perception, neutrality and nonalignment are not identical. But the difference is not obvious. While neutrality is a time-honored classical term of both international law and political history, nonalignment is a new concept, first used in the 1950s to designate countries that refused to be aligned with the political-military alliances of either the Western powers or the Soviet Union in the worldwide confrontation between West and East. Nonaligned did not, and does not today, mean politically or even militarily neutral in an international conflict. A nonaligned country like Libya, Syria, Iraq, or India may well take sides in a war and become the military or at least the political ally of another power. Nonalignment is a notion of refusal, historically linked to the formation of the blocs or groups of nations and the East-West conflict.

Switzerland: The Precedent

The neutrality of Switzerland has a unique character. Its military and political significance is determined by history and geography as well as by an economic foundation and demographic structure of national defense, unique in Europe and therefore only roughly comparable to other countries. It is less the militia force structure that characterizes Swiss armed neutrality than it is the dimension of the Swiss people's national defense and the military prowess of the Swiss army as well as the national consensus on military defense and its requirements in peace and war. Swiss neutrality was agreed upon by the European powers at a time when European wars were still to be expected as recurring continental events—the historical experience since 1815 proving this point. Swiss armed neutrality held during the entire nineteenth century after the Congress of Vienna and stood the test during the World Wars I and II when Switzerland, bordering on France and Germany, maintained herself neutral but ready to defend her borders. The warring powers respected this neutrality. In the case of France this was certainly to her own advantage, and it had been so considered by the French general staff in 1914 and in 1940—apart from the policy, which excluded any design on Switzerland. While this applies equally to Germany in 1914–18 as well as to the German coalition in 1870–71, the case was different for Hitler in 1940–43. The strategic situation after the defeat of

France and the control of northern Italy until shortly before Germany's surrender made it unnecessary for German military strategy to commit forces for the occupation of Switzerland. The political advantage of having one neutral country in the middle of Europe was not to be underestimated by either side in either war.

This experience points to the dominant interest of the Western countries regarding Swiss neutrality. Switzerland, capable of effective national defense in a conventional war around her borders, provides a considerable indirect deterrent in military terms against an offensive war in the two strategic directions from eastern-central Europe: to the west and south. This defensive deterrent tends to support the stability of the European situation and backs up to some degree Austria's neutral position. Swiss neutrality and security are in turn supported by a strong NATO defense in Germany and northern Italy. The Swiss position would be covered more effectively by a more capable Austrian national defense, as would the strategic defense of Yugoslavia.

Swiss armed neutrality, based on the national defense posture of a modern conventional mobilization army and a capable air defense, can be considered a stronghold of European security because of its potential stabilizing effect in a crisis situation. The geographical situation of Switzerland permits timely mobilization and preparation for active defense as long as NATO forces hold their own against an attack from eastern Europe in Germany. However, independent deterrence cannot be expected from Switzerland in the case of a European conflict with a threat of general war.

But as with Sweden's, Finland's, and Yugoslavia's defensive postures of armed neutrality—in the Finnish case with more severe limitations—Swiss armed neutrality is to be considered an asset for the security of Europe under the prevailing geopolitical conditions.

The Cases of Sweden and Finland

Both Nordic countries are neutral in the sense that they are not members of a multilateral military alliance. However, their respective situations are different: Finland is bound to the Soviet Union by the Soviet-Finnish treaty of mutual assistance, concluded in 1948, which obliges Finland to fight aggression against the Soviet Union on Finnish territory with all her military forces if the Soviet Union is attacked by Germany or any other country allied with Germany.

While this treaty of the postwar period is based on a German attack, the Soviet Union could interpret Finland's obligation as pertaining to a contingency which involved any German state and its allies—i.e., NATO since 1955. This is certainly a purely hypothetical contingency, given the strategic configuration of Europe and the nature of the North Atlantic alliance as well as of the Federal Republic of Germany. But the Soviet Union retains a measure of coercive legal power over Finland by virtue of this treaty. Finland's geographical situation and the balance of forces in northern Europe and the Baltic, with Soviet predominance, are compelling regarding the exercise of Finnish national defense and foreign policy in a crisis. Therefore, Finnish neutrality is limited in favor of the Soviet Union by fact as much as by legal obligation and by national interest.

This context does not change, however, Finnish preferences and the national interest to maintain independence, territorial integrity and—if at all possible—armed neutrality in a conflict between the Soviet Union and Western powers or NATO as a whole. Since an attack by NATO against Finland or the Soviet Union through Finnish territory is inconceivable, the treaty contingency cannot materialize: the NATO allies would not initiate armed hostilities against the Soviet Union or the Warsaw Pact, and NATO forces could not enter Finnish territory until after having marched through either Soviet or Swedish territory. For both reasons, there is no real threat to either Finland or the Soviet Union.

With this political mortgage on its independence and neutrality, Finland has managed to stay outside the Soviet empire and to consolidate her internal integrity after the postwar crisis. By all standards Finnish neutrality in an East-West conflict can be considered real and loyal to all countries. "Finlandization" by Soviet pressure and subversive influence may be a hypothesis of political conjecture; it is certainly not a reality today nor has it been an expression of Finland's internal condition during the past three decades. Finlandization as an effect of Soviet policy or of Finnish adaptation to Soviet power is not in evidence. In fact, Finland's cautious attitude toward the Soviet Union, and her critical attitude regarding certain aspects of NATO policy in northern Europe, translate the geopolitical situation of the country and the requirements of Finnish national policy, though not its accepted dependence on the Soviet Union. A balance between East and West is a critical part of Finnish foreign policy.

Finland's delicate situation depends on the Nordic balance, itself a fragile state of affairs. The basis for Finnish self-assertion is the armed neutrality of Sweden and the Swedish capability to offer resistance to any military aggression. As long as Sweden remains truly neutral in a crisis and is able to defend its neutrality and independence, backed up by her strategic-political environment to the west and south, Finland has a fair chance of maintaining herself even in the midst of a European crisis. More cannot be asked of either Finland or Sweden in such a contingency.

Sweden's independence, territorial integrity, national defense, and neutrality rest on the Finnish glacis to the east and the North Atlantic alliance to the west and southwest, i.e., on Norway, the western control of the Danish Straits and the western part of the Baltic Sea, at least. This is not to say that Swedish armed neutrality is entirely dependent on NATO. But there is no doubt that the strategic situation of Sweden depends on the wider framework of the strategic environment in Scandinavia and the Baltic. This environment is determined by the political divide of the Baltic to the advantage of the Warsaw Pact, with most of Sweden's shoreline on the eastern side and opposite Finland. Soviet submarine explorations within Swedish territorial waters are operations of opportunity as well as preparations for hypothetical contingencies in which the Soviet navy and amphibious forces would carry Soviet forward defense close to the Swedish coast in order to use Swedish territory as a western glacis or even as a territorial base for an offensive against Norway.

This situation, with its possibilities for the Soviet Union, places a considerable defense burden on Sweden if neutrality is to remain real in times of crisis. It also requires a balanced foreign policy between East and West in all matters pertaining to European security, in particular in Scandinavia and in the Baltic. But it is dependent for its own balance on the deterrent power and defensive preparedness of the North Atlantic alliance to the West—in practical terms, on the military solidity of NATO's northern flank. This in turn hinges on the capability and political reliability of the United States and Britain as Norway's allies and the forces they could deploy to the Norwegian Sea and to Norway in time of need. In this context German and Danish contributions for the Western control of the Danish Straits are essential to Sweden's security.

Politically Sweden is to be considered part of the West. Swedish foreign policy cannot but express this essential feature of Sweden's nature. There could be the risk, in certain situations, of a conflict between the necessity of a balanced foreign policy for the sake of protecting neutrality and the preference for a political orientation favorable to the West.

Such situations could also produce strains on Sweden's relations with the East or the West, according to circumstances and to the political challenge posed to the principles and objectives of Swedish national policy. Such was the case for Swedish-American relations in the 1960s during the Vietnam War. A more pertinent and direct case in point was the tension during the Polish crisis of 1980–81, when Sweden anticipated the repression of the Polish revolutionary labor movement under Soviet pressure, which could have provoked a state of active hostilities on the western and northern borders of Poland and a mass flight of Poles to the west across the Baltic.

Such contingencies could arise at any time in the future if the brittleness of communist regimes in Eastern Europe and the tenuousness of Soviet political control are tested by the social forces under pressure of crisis, be it of political or economic origin. Under such circumstances Sweden, as a free and democratic country of the West, will always be confronted with a challenge and may be exposed to danger. Thus Swedish security, and with it the reality of Swedish neutrality in conflict, are linked to the development of a conflict between NATO and the Warsaw Pact; that is, their maintenance depends on the crisis stability of the strategic East-West relationship, involving the two superpowers and their European allies.

Austria

Austria experienced the risk of involvement in acute military conflict twice since her newly won independence: in 1956 during the Soviet intervention in Hungary and in 1968 during the Soviet intervention in Czechoslovakia. Both cases were inconclusive, since neither crisis produced a serious European contingency with an actual risk of war. Hence, Austrian security was not really at risk, and Austrian neutrality was not put to the test. But both events point to the unstable geopolitical situation in central Europe; Austria's political position between East and West could be endangered by a military threat caused by events in Soviet-controlled Eastern Europe.

It can even be argued that both the Hungarian revolution of 1956 and the events of the political Prague Spring in 1968—with the ensuing Soviet military interventions and prolonged occupation of both countries—were partly caused by these countries' common borders with Austria, a free and neutral country of the West. Austria, of course, also must be looked at as a border country with relatively easy access from the north and east, i.e., from Bohemia, Slovakia, and Hungary, and as a territory of approach to Western Germany, Italy, and northwestern Yugoslavia for WP forces in an offensive war against the West. The options offered by Austria's geographical situation for an offensive war by the Warsaw Pact are more obvious and promising than those which a Western coalition or power could use against east-central Europe. Given the balance of forces and the force structures and deployments of both alliances in central Europe, the opportunities for the Warsaw Pact would seem largely greater than those for NATO, even for a second phase of war counterattack. This situation means that Austria's neutrality—like that of Finland, Sweden, or Yugoslavia—would be more threatened in a European crisis from the East than from the West.

However, Austrian neutrality, according to the treaty status, must be passive and balanced with regard to each side; i.e., it must be based on hypothetical contingencies involving a Western as well as an Eastern threat to Austria. This highly artificial foundation of Austrian security and neutrality is part of the geopolitical foundation of Austria's unity and independence. The power politics of the European postwar period are perpetuated in this historical legacy: the imposed quality of Austrian neutrality.

Austria's neutrality must be defended within treaty limits, which restrict military forces and capabilities. Hence Austria does not have the political and legal possibilities which characterize Sweden's, Yugoslavia's, and Switzerland's armed neutrality, or that of Spain before it became a member of the North Atlantic alliance in 1982. Austrian neutrality, therefore, is less independent and it is militarily more threatened by geographical exposure to attack.

In the interest of crisis stability in central Europe, and therefore in the general interest of European security, Austria ought to be free to defend her own security and hence her neutrality by the necessary military means within the limits of her economic-financial and manpower resources. The narrowness of the territory and the airspace add to the problem of national defense. The military answers defined by Austrian national defense are not the object of this argument. The Austrian national defense posture tries to make do with what is possible under the imposed conditions of neutrality.

The Case of Yugoslavia

Yugoslavia presents perhaps the most interesting and significant example of nonaligned policies in Europe, while its neutrality seems to be more questionable than that of any of the other neutral countries in Europe.

President Tito was the cofounder of the international movement of nonalignment. He based Yugoslav foreign policy on this foundation, which permits international activity, including participation in political operations of one group of countries against another, and which offered Yugoslav communism the opportunity for an anti-imperialist stance as required by ideology. While Nehru and Nasser had defined neutralism as an active force and nonalignment as a policy of self-assertion against the supremacy and hegemonial tendencies of the great powers of the West and East, Tito tried to build a position of self-sufficiency in Europe, independent of either alliance, secure between East and West, and capable of serving as a solid base for active international policies, including support for political forces and countries in a revolutionary anti-imperialist struggle against colonial or postcolonial foreign domination. This external aspect of "Titoism," which still inspires Yugoslav foreign policy, has to be considered when defining Yugoslavia's special brand of neutrality and nonalignment. For example, Yugoslavia has supported, politically and with supplies of arms and matériel, the Greek-Cypriot fight for enosis, and the fight of the Palestinian Arabs for unity and independence of Palestine. Yugoslavia is still in league with certain Arab underground combat organizations and active in the Third World.

In European politics Yugoslavia holds a particular and ambiguous position between East and West. Politically the country belongs neither to the communist-ruled East nor to the democratic-liberal West. The internal conditions still resemble more those in socialist Eastern Europe than those in Western Europe: the regime is still the dictatorship of a class party, and authoritarian-bureaucratic socialism has only been moderated by the "workers' self-administration." But this moderation has had a growing effect on the economy, social order, and internal development. Although Yugoslavia's frontiers to Western Europe are not quite open in the Western sense, they are not closed according to the Eastern example; even in the Eastern direction it is possible to access and exit. Yugoslavia is a country of transition between East and West and is more open than all East European countries, even Hungary.

Whatever the internal political reasons for this situation—which has caused many contradictions and internal tensions—it has been primarily caused by Yugoslavia's economic dependence on trade with the West and the fact that Western financial credit has become the basis for the country's independence. The workers' interest in a better life cannot be met by the socialist cooperative economy of Yugoslavia. Therefore the migration of part of the labor force to industrialized countries in Western Europe is a necessity as well as a political safety valve for the regime. In this respect Yugoslavia resembles Turkey.

This migration, as well as the old ties of Slovenia and Croatia (both Catholic and of Latin culture with Austrian influences) to the neighboring Western countries Austria and Italy, make Yugoslavia dependent to a large extent on secure and cooperative relations with Western Europe. One can ask whether the inclination of nineteenth century Serbia toward France and the legacy of the entente relationship between the two world wars exercise a balancing influence on the conception of modern Yugoslav European policy. The answer can only be tentative. The essential inspiration remains Tito's legacy: independent national communism based on national independence and federal unity of the various ethnic parts of Yugoslavia.

Yugoslav international activities and foreign policy have not, so far, alarmed the North Atlantic alliance. Although Yugoslavia has, on occasion, joined in the call by other Mediterranean countries for a denuclearization and even demilitarization of the Mediterranean Sea in the sense of a withdrawal of both Soviet and U.S. naval forces, Belgrade has not pushed this issue. Occasional Soviet and American warship visits to Yugoslav ports have been balanced. The evidence suggests that while Yugoslavia cannot contradict the demands of other countries, it has, for the time being, a dominant interest in maintaining the status quo, since the presence of U.S. forces in Italy, Greece, and the Mediterranean is a relative guarantee for the security of Yugoslavia in a contingency created by a Soviet offensive in southeastern Europe or the Mediterranean.

Yugoslavia is a member of the theoretical Balkans security pact with Turkey and Greece—a historical legacy from the prewar period. In 1934 the Balkans Pact was enacted between Yugoslavia, Greece, Turkey, and Romania against Bulgarian territorial revisionism. In 1953 a new friendship treaty was concluded between Yugoslavia, Greece, and Turkey after the change of Yugoslav attitudes toward the West, and in particular toward Greece, brought about by Tito's conflict with the Soviet Union and his new independent foreign policy based on a neutral position in the East-West confrontation in Europe in 1948. The 1953 treaty envisaged cooperation for common regional security including common defense in case of an unprovoked attack. Greece and Turkey were already members of the North Atlantic alliance and its military-political organization NATO since 1952. On this basis the new Balkans Pact was concluded between the three countries in 1954 for twenty years with the provision of silent renewal. This treaty for collective defense with mutual assistance is still valid. It has, however, no real significance, since it was never translated into a practical security agreement, let alone into a military alliance.

The Western powers included Yugoslavia in the Western

sphere of security when, in February 1951, during a state of tension in the Balkans, the United States, Great Britain, and France declared that they would consider an attack on Yugoslavia an act of aggression. Accordingly U.S. President Lyndon Johnson warned the Soviet Union publicly, during the Czechoslovakian crisis of the summer of 1968, not to threaten Romania and Yugoslavia. It can be concluded that Yugoslavia's security and neutral status are essential features of European security in the Western perception. In fact, an extension of the Warsaw Pact to the Adriatic coast or the occupation of Yugoslavia by Warsaw Pact forces would critically change the strategic situation of NATO's southern command and southeastern flank as well as Italy's security equation.

Thus Yugoslavia and NATO share an interest in maintaining the political-strategic status quo by mutual support. Yugoslavia's national defense strategy of "people's defense," with general mobilization, conventional forces, and guerrilla operations by regional and local militia combat groups, could put up a stiff resistance to invasion but would not be a match for long for a massive Soviet force. However, the prospect of having to fight in order to occupy and use the country, and the risk of provoking counteraction by the West can be considered to be sufficient deterrence to protect Yugoslav neutrality and independence.

NATO's interest certainly recommends the support of a Yugoslav fight for independence if that country were threatened by the Soviet Union. It is obvious that a strong national defense of Yugoslavia is an asset to Western security, as long as Yugoslavia remains truly neutral regarding the Warsaw Pact. NATO military options in such a contingency as mentioned above, and envisaged by the United States in 1968, would be limited. However an attack on Yugoslavia hardly seems likely as an isolated event with the rest of Europe at peace. Therefore, such a contingency must be considered as part of a larger one, involving Europe as a whole. In a general European conflict neutrality of uncommitted countries would become a serious problem—and in this sense the Yugoslav case would merge strategically and politically with the general European situation created by large-scale aggression.

THE SIGNIFICANCE OF ARMED NEUTRALITY
FOR NATO STRATEGY

As revealed above, NATO assumes that the neutral countries of Europe would defend their frontiers as best they could under threat of aggression or actual attack. If this assumption proved wrong, one of two different situations could occur: the neutral country that was attacked or occupied because its resistance was ineffective would be treated by the aggressor as hostile and would collaborate only under military coercion; or the country would willingly comply with the aggressor's demands, supporting his war effort against other countries by active or indirect military assistance and by allowing his forces to use its territory as a logistical or operational base for the aggressor's offensive. In the first scenario the Western allies probably would not consider the conquered/occupied country part of the Warsaw Pact, even if it had become engulfed by the WP strategic-geopolitical sphere. NATO might well, then, spare that country as much damage and fighting as possible and even try to support the population and any active resistance there (as was done between 1941 and 1944 in the cases of Greece and Yugoslavia, both of which, however, had joined the Allied camp). In the second scenario such a formerly neutral country, having joined the adversary's war effort for all practical purposes, would eventually have to be treated as a hostile country and would be dealt with according to NATO strategy and to NATO's best advantage.

In the face of war (specifically, an international crisis), armed neutrality can resemble a color scheme in various shades of gray: ambiguity characterizes the posture of the neutral country, who has the task of managing both fighting sides in order to avoid attack—unless, of course, the neutral country is so powerful militarily that it can remain isolated.

This is not the case in Europe today nor was it in either the First or the Second World War. NATO policy and strategy will take the unavoidable ambiguity of neutrality under certain circumstances into consideration, just as the Allied powers did in most cases in 1939–45 with regard to neutral European countries. This does not, however, provide assurance in advance to countries that do not intend to defend their neutrality by opposing aggression on their borders, including those with a policy and military preparedness meant to discourage aggression.

Seen from the West, there are essentially three different geopolitical-strategic situations on the border regions of the North Atlantic alliance in Europe, as discussed in the first part of this chapter.

1. Finland and Sweden on the northern flank of NATO
2. Austria in central Europe
3. Yugoslavia in southeastern Europe

As explained above, NATO military strategy counts on the northern flank when planning for the possibility of a Soviet offensive in the Baltic toward the Danish Straits and Sleswig-Holstein (a), as well as for a Soviet offensive through Finland and Sweden toward northern Norway in conjunction with a direct attack from Kola by air, sea, and over land (b). Both contingencies could be combined in a general war against NATO, or the second option could be exercised by the USSR as a limited offensive aimed at securing a forward position in northern Europe and the Norwegian Sea during an international conflict short of general war.

The first contingency (a) is extremely unlikely to materialize as a separate and limited offensive action by WP forces: such an attack would constitute a severe threat to the security of NATO forces in Germany and to the security of central Europe in general. All the nuclear powers of the Western alliance have forces in Germany and have a political prime interest to defend there. The second contingency (b) has been considered a distinct possibility for a long time, i.e., as a separately viable and valuable offensive option of Soviet strategy in an international crisis as much as in war. The probabilities of either contingency will not be discussed here; it can be assumed that both are in reality rather low. But both could materialize at the outset of a general war by the WP against NATO. NATO strategy must be prepared for either.

The first contingency would be more likely in the case of an all-out attack by WP forces against NATO; it would be part of a general offensive war against Western Europe. The object of WP operations in the Baltic would be a main thrust against the central sector and across Germany in an attempt to open the Danish Straits and later to seize southern Norway for forward deployments of naval and air forces against NATO's North Sea defense, West Germany, the Channel, and Britain. Sweden and Finland could be touched by such WP offensive operations, but would not become major objectives. NATO, therefore, would not need to commit forces to cover these countries or to actively deny access to them by interdiction. A WP breakthrough attack on the Danish Straits would probably infringe on Sweden to some extent, but its objective would lie to the west of Sweden. NATO would need all its available forces in the region, including U.S. and British reinforcements, to deal with such a situation; it could spare no cover for Sweden in such a contingency.

The second contingency in the high north would have to include both Finland and Sweden if the ultimate goals of an attack by Soviet forces from Kola on Norway were the occupation of north Norwegian airfields and ports and the forward deployment of the Soviet Northern Fleet west into the Norwegian Sea under land-based air cover. This contingency is the standard scenario for a military threat to northern Europe by the Soviet Union. In this case, northern Sweden would have to be used by the advancing Soviet ground forces coming from Finland. These forces would have to be deployed from the Soviet Leningrad military district through Finland and supported by air attacks against Finnish and Swedish resistance. This would mean a considerable war effort on the Soviet side, demanding reinforcements from the operational and strategic army and air force reserves in the European USSR. Therefore NATO hopes Sweden and Finland will demonstrate their readiness for national defense of their neutrality in any acute crisis in order to deter the Soviet Union from such an offensive.

If both Finland and Sweden would defend themselves against actual attack on their borders, then NATO could rely on a retarding action and attrition of Soviet forces in Scandinavia, offering precious time for the reinforcement of Norway and the forward deployment of U.S. and British naval forces to the Norwegian Sea. Furthermore, Soviet forces would be stalled in the north, far away from the main battleground of initial attack against Western Europe.

The prepared Swedish territorial defense in northern Sweden

and the Swedish air defense are therefore considered assets to the security of Norway in terms of time and space. The defensive options of Sweden against such an attack from the East are considered to be real and significant, as are those of Norwegian defense. Armed neutrality and effective Swedish and Finnish national defense are therefore perceived by NATO as support for the defensive posture of Norway and hence as assets for NATO's defensive strategy of denial by flexible response.

One must keep in mind that the Soviet Union has placed a two-edged sword on top of Scandinavia: the Northern Fleet and its land-based aviation is an instrument of offensive war to be used early in a conflict in order to gain the eastern North Atlantic approaches and seal off the Norwegian Sea to NATO naval forces and reinforcements for Norway. But the Kola base complex is the most massive strategic forces concentration for the second-strike capability of the Soviet Union. This base complex is within reach of conventionally armed tactical aircraft based in northern Norway or taking off from aircraft carriers in the Norwegian Sea, let alone the nuclear options of U.S. and British air and naval forces.

Soviet strategy since the 1950s has tried to keep U.S. and other NATO naval forces with carrier-based attack aircraft as far away as possible from the Kola peninsula and the North Sea frontier of the Soviet Union, the Barents Seaboard, in general.

While it would be in the operational interest of the Soviet Northern Fleet and its land-based aviation, as well as of the strategic ballistic missile submarines based on Kola, to move out into the North Atlantic, or at least for the SS-N-18 and SS-N-20 boats to deploy into the Barents Sea, and while it would be of considerable strategic-operational advantage to the Soviet forces to forward deploy into the Norwegian Sea in order to deny access to Western forces in time, such forward movements could well precipitate events and draw upon the Kola complex precisely the kind of attack at the beginning of a war that such offensive operations would be launched to prevent.

Both Soviet strategy and NATO strategy in northern Europe thus are confronted with considerable risks in any case and the situation might become unstable at once. The preemptive attack options in the north represent the destabilizing tendency in the East-West balance of military forces and options on the northern flank of NATO and the northwestern flank of the USSR. Therefore, Sweden's and Finland's armed neutrality can compensate somewhat for this tendency and lend added stability to the situation in crisis. Seen from this angle, NATO strategy has a major interest in maintaining the neutrality of both Sweden and Finland as long as possible, using the time their defense efforts would offer to stabilize the northern flank.

Austria in central Europe presents a different case, as discussed earlier, because of its narrowness and exposure as well as its central location between southern Germany and northern Italy. NATO territory and forces in the FRG can be attacked from Austria south of the Böhmerwald/Bayerische Wald by WP forces coming from Bohemia south into northern Austria and turning west in the general direction of Munich, thus avoiding the obstacles of the mountains and outflanking the prepared defenses of NATO forces in Bavaria. Such forces would not enter the Alps but stay and move in the Danube region. Austrian defense would have to hold or at least delay them there. Hence, the importance of a sturdy Austrian national defense to both NATO and Switzerland. This contingency has preoccupied NATO ever since 1955. NATO forces in Germany are too dispersed for forward defense and lack in operational reserves (hence the importance of French forces to back them up in time). A WP offensive through this part of Austria would create a critical situation for NATO defense. In this context the air defense problem of Austria takes on very serious proportions not only for Austria herself but for the FRG, Switzerland, Italy, and for NATO forces in Bavaria.

In the east WP forces would not need to go south from Vienna in the direction of Lubljana and Zagreb or Trieste, since in Hungary they are nearer to Yugoslavia and northern Italy. But they could be supported east of the Alps along a north-south axis in their advance from Hungary into Slovenia, Croatia, and Istria to turn on northern Italy. In this respect, the air defense issue of Austrian neutrality would seem to be even more critical than the ground defense problem.

On the basis of this assessment, it is obvious which side profits from Austria's inability to defend her airspace and how fragile Austria's neutrality could become under Eastern pressure in an acute European crisis. NATO strategy has but one interest with regard to Austria: that Austrian neutrality be defensible and remain stable in a crisis.

Yugoslavia, like Finland and Sweden, offers a territorial barrier between the Warsaw Pact and NATO. Yugoslav armed neutrality lends stability to NATO's southeastern flank just as NATO lends stability to Yugoslav national defense and hence neutrality. The mutual interest in this geopolitical-strategic situation is obvious and unquestionable. This role for Yugoslavia is particularly notable in the cases of Italy, across the Adriatic Sea, and Greece, in the south. For this reason Tito was never interested—since his separation from the Soviet bloc at the end of the Greek Civil War, after he had withdrawn his support for the Greek Communists in 1948—in any scheme of Greek politics to establish nuclear-free zones around Yugoslavia or for Greece to leave the military cooperation of NATO. During his last years he discreetly tried to dissuade Greece from any experiment in neutrality, since this would, of necessity, weaken the strategic position of Yugoslavia, exposing it to attack from the north and east. It goes without saying that Yugoslavia, with her particular neutralist foreign policy and international commitment to the cause of the nonaligned movement, must not appear to be in favor of U.S. presence in the Mediterranean, of U.S. nuclear weapons in Europe, and of U.S. bases in Greece. But there is a difference in this case (as in many others) between official "declaratory policy" and the real strategic interest of a country in a difficult geopolitical situation. NATO strategy cannot build additional security for either Italy or Greece on the independent and neutral position of Yugoslavia; but it can use the given situation, which offers time and space in case of a general attack by the WP against Western Europe.

The Yugoslav contingency that would threaten Western security in an international crisis has generally been described as a falling apart of Yugoslavia under the internal strains of divergent interests and the WP stepping in to either salvage Yugoslav federal unity to communist and Soviet advantage or, on the contrary, to carve up Yugoslavia in favor of old territorial claims, suppressed since 1945, by Hungary and Bulgaria. The realism of this proposition and the likelihood of the occurrence of such a political situation, offering such opportunities to the Soviet Union and its allies as well as challenges and risks to their own interests, is a matter of conjecture and need not be discussed here. But the possibilities must be considered by NATO policy and strategy.

It is obvious that NATO could not allow the WP forces to camp on the coast opposite Italy, a distance of some 200 to 300 km across the Adriatic Sea, and to use the airfields there, without losing what security there is for Italy. The geography of the Adriatic region spells vulnerability for Italy from Trieste to the Strait of Otranto. Had Yugoslavia joined the Warsaw Pact in 1955, and had it been integrated into the WP forces along with other countries under Soviet command and control (and provided with standardized Soviet equipment for the air force and air defense), the entire situation would have been changed to Eastern advantage. This applies not only to the Adriatic and northern Italy, but also to the U.S. Sixth Fleet, based in Italy for a large part, and the command of the eastern Mediterranean. In the event of war, NATO would have been immediately confronted with an active theater of operations. The Greek situation would have become untenable in the case of an attack on a Macedonian front, considering the short distance to Salonika. Greek defense would have been much more dependent on the prompt landing of Western expeditionary forces, on the backup and supply from the sea, and on NATO air forces. The Turkish Straits would have been permanently outflanked by standing WP forces; their defense would have become even more difficult than it is already, given the facts that Bulgaria shares a border with Greece and has little land behind it to the coast. Albania would have remained in the WP and under Soviet control. The entire geopolitical-strategic environment would have been changed to the advantage of the Soviet Union as long as Moscow remained in command of Yugoslavia.

To prevent such a change from taking place has been one of the major political-strategic objectives of NATO. Thus the value of Yugoslav armed neutrality and a capable national defense against invasion on the Balkans can hardly be overestimated. As set forth above, and earlier in this chapter, NATO is compelled to maintain the present situation, and in war, to deny the WP the military use of Yugoslav territory and military installations in Italy and Greece. To interdict the use of the airfields and to close the Strait of Otranto would be an imposing necessity in the event of a WP occupation of Yugoslavia.

Expert opinion in the West describes such contingencies as involving a scenario in which a group of Yugoslav communists, established as a national and socialist political authority, invite Soviet forces to come to fraternal help. This consideration is, of course, linked to and even based on the 1968 Prague example and the implications of the Breshnev doctrine of the superior interests of the "commonwealth of Socialist" peoples, the common security of which takes priority over national interest.

Whether a counterintervention could be launched or whether the WP's use of Yugoslavia's military assets against NATO could be effectively interdicted are matters of the risk of war involved and the means of action available. The prospects of a military confrontation, with its risk of nuclear escalation, over the violation of another country's neutrality and independence (and, for that matter, one outside the Western alliance) would be hard to face by any Western government. Therefore, given the balance of forces and the limitations on NATO assets in southern Europe, NATO's strategic options would have to be very carefully managed in such a contingency.

Again, the question must be asked, is a separate Balkans war contingency, provoked by the Warsaw Pact, likely? Moscow's willingness to take a risk for such a purpose must be assessed in the West. Ideology and limited political-territorial gains would not seem to be sufficient reason for the Soviet Union to risk war with the United States over Yugoslavia. However, in the summer of 1968, the U.S. government could not exclude such a possibility. This should serve as a reminder that crisis can escalate into conflict, once the train of events has begun. Control of such a development by Western moderation could well lead again to Soviet net advantage—the gaining of control over Yugoslavia—thus changing the situation in southern and southeastern Europe. Acquiescence to a fait accompli of such magnitude would certainly weaken the Western alliance, since it would undermine the U.S. position in Europe, Italy, Greece, and Turkey in particular, and would destabilize NATO's southern flank as well as European security as a whole. It would not do, on the other hand, to make preemptive gestures about the security and neutrality of Yugoslavia and to create the impression that Yugoslavia were somehow part of the Western security complex, covered and claimed by NATO and the United States. Therefore, the situation has all the makings of a dilemma, difficult to solve under the strain of an acute crisis.

The cases of neutrality discussed here, with respect to NATO strategy in crisis and war, have several features in common, all of which point to the extraordinary complexity and ambiguity of the European neutrals' place in the geopolitical-strategic equation of European security. These features include

1. limitations of effective national defense of independence and neutrality during actual conflict;

2. the importance of the cohesion and effectiveness of the North Atlantic alliance as a deterrent to aggression and military pressure against neutral countries in Europe;

3. the change of the geopolitical-strategic situation between East and West, to the detriment of both the neutrals and NATO, to be expected by successful offensive strategy and expansive policies of the Soviet Union;

4. reliance on the stabilizing influence of American military presence;

5. the necessary balance of policies between the two sides;

6. preoccupation with the risk of escalation of conflict and with U.S. military presence, nuclear arms, and policies of crisis management;

7. the resulting ambiguity, both of neutrality and of balanced foreign policy, in neutral countries;

8. overstatement of the value and safety of the status of neutrality in those European countries that belong to one of the two alliances, and hence the temptation of neutrality.

This last feature became evident between 1981 and 1986 in Spain during the long-lasting debate about membership in the North

Atlantic alliance as it has emerged in other NATO countries in the last years. The risks and burdens of neutrality have often been overshadowed or diluted by the fears of war and involvement in conflict in those countries, where U.S. forces and nuclear arms are deployed, where U.S. forces would arrive in time of crisis, and where large NATO military installations are situated. This feature was even present in de Gaulle's argument for the advantages of national deterrence prior to, during, and after France's withdrawal from the military system of NATO in 1966–67. It is part of the makeup of Greek politics and policies and is present in various other Western European countries, as could be seen in 1981–83 during the campaign for or against the deployment of U.S. LRINF missiles in Europe.

This political factor has to be considered when the merits and risks of both neutrality and membership in a military alliance are being discussed. For NATO strategy this feature is a political factor in terms of the acceptance of alliance politics and NATO strategy.

CHAPTER SEVEN

THE NEUTRALS,
THE SOVIET UNION,
AND THE WEST

Curt Gasteyger

On a somber day in October 1944 Marshal Stalin discussed with his British guest, Winston Churchill, the vagaries of war and the prospect of peace. As regards the first, Stalin proposed, among others, an Allied operation across Switzerland. Its purpose would be to circumvent the Siegfried Line, which then seemed to be a formidable obstacle to a further Allied advance against Germany. He also planned to draw the Swiss army into the fight against the Germans. In the end Churchill succeeded in persuading Stalin that such an operation would be counterproductive and should be forgotten.

This episode finds its reflection in an instruction which Churchill later gave to his foreign secretary, Anthony Eden, dated 3 December 1944.

> "I was astonished at U.J.'s [Stalin's] savageness against [Switzerland], and, much though I respect that great and good man, I was entirely uninfluenced by his attitude. He called them "swine," and he does not use that sort of language without meaning it. I am sure we ought to stand by Switzerland, and we ought to explain to U.J. why it is we do so."[1]

THE AMBIVALENCE OF NEUTRALITY
AND ITS UTILITY

This story shows that the Soviet leader had little patience with Swiss neutrality. To him, neutrality in war was politically suspicious and strategically expendable, particularly if practiced by a "capitalist" country. That Churchill in the end succeeded in pursuading the Soviet dictator not to involve Switzerland in the war was, of course, a relief for the latter. But it does not necessarily prove that Churchill, or indeed the United States, had particular sympathy for neutrality either. When describing the measures of economic warfare against the neutrals, the then assistant secretary of state, Dean Acheson, reported: "At home the public, almost to a man, regarded arrangements to supply the neutrals as traitorous connivance at treating with the enemy. Neutrals were judged to be enemy sympathizers."[2]

Little love was thus lost for the neutrals during, and for many years after, the Second World War. The final breakdown of the European system and the dawning of a totally new kind of conflict—that between communism and Western democracy—raised in many quarters the question of whether neutrality could or should still have a future. To many, it seemed to be little less than a folkloric leftover of bygone times, of a system of national states that had totally discredited itself in the war and had also little justification in the emerging cold war. In times of confrontation statesmen and people find it difficult to distinguish between political neutrality and moral indifference, between the neutral's desire for independence and its duty for demonstrating solidarity. In such times there can be no neutrality between "good" and "evil" but only a closing of ranks to defend the former against the latter. But while in Western democracies such attitudes toward the neutrals are rarely, if ever, a matter of principle and more an expression of mood, in the Soviet Union they were an outgrowth of ideological doctrine and as such part of official policy. To phrase it differently or more concretely: the Soviet Union looks at neutrality as a means and not as an end. To her, neutrality is impossible by any standards of ideology. It may be acceptable politically, but only if it is seen as an intermediary phase on the country's way toward communism.

THE ROLE OF THE NEUTRALS IN EUROPE

In the early years after 1945 the central question for the neutrals was that of their future place and role in postwar Europe. In a sense the question was similar for all small countries: the age of truly global powers and their ever-widening competition had begun. Under such circumstances the national state appeared to have outlived itself and so, it seemed, the idea of neutrality.

In Western Europe plans and projects for economic and political integration flourished, and in the East the Soviet Union tried to mold national identities into a uniform communist bloc under her undisputed leadership. Thus little, if any, room seemed left for small entities, be they neutral or allies, "capitalist" or "communist."

This situation of uncertainty and doubt is well described in a recent article.

> "Some forty years ago, when the present international system was emerging, it was a widespread opinion that in the future the small countries would find themselves increasingly powerless and unimportant. The states about to become known as Superpowers were superior when it came to the resources which were considered to be crucial, and the power gap was predicted to widen during the decades to come."[3]

That was the mood in many quarters of Europe: the visions went toward greater units, more integration, and, with the mounting menace of another war, collective or regional security in the form of newly created alliances. But in spite of all this the nation-state showed either greater resilience or simply more inertia than was anticipated. The dream of a (West) European federation receded into the background; Stalin's efforts to forge Eastern Europe into gray uniformity stopped short of abolishing the nation-state. The trend reversed: instead of disappearing, the nation-state in the fifties revived and reasserted itself. This was facilitated by a gradual mellowing of the East-West confrontation and timid beginnings of various kinds of dialogue across the political and ideological barriers that separated—and still separate—Western from Eastern Europe.

The neutral states were at the same time victims and beneficiaries of these fluctuations. Hardly liked for their "abstentionism," first during World War II and then in the early days of the cold war, they gradually regained some respectability when their contribution to European stability and their utility in promoting the East-West dialogue was (sometimes reluctantly) recognized. This took some time and some doing on their part. It also took some adjustment on the part of the allied or non-neutral states.

Unlike the newcomer to the international scene, the Soviet Union, Western countries had long learned to live with neutral states. After all, in almost every aspect of daily life, except foreign policy, the neutrals had always been part of what after the war came to be known as the West. They were pluralist democracies with a high and growing degree of industrialization and economic interaction. In spite of their neutrality they were considered, in 1950 as well as today, by both the Western alliance and the Soviet Union, to be sitting clearly on the Western side of the East-West divide. In doing so they became an integral part of Europe's postwar system and are now fully accepted as such.

This is not to say that they all practice identical policies of neutrality. As the various case studies show, these policies differ in several respects. Neutrality has more than one father. Its nature and scope are a function not only of tradition but also of location, of interpretation as well as of specific interests. Thus Sweden's neutrality acquired additional importance when the Soviet Union hesitated to tolerate Finland's neutrality. She would hardly have done so in the long run had Sweden joined, as some people in the West hoped, the North Atlantic alliance. At least to the Soviet Union, it was important to have a neutral Sweden as Finland's western neighbor. But Finland, too, probably feels more at ease in company of another neutral in the region than if she had to face two military alliances.

Switzerland's case is somewhat different. Having survived World War II she did not hesitate to maintain neutrality, which had served her so well in the war. But unlike Sweden she did not become a member of the United Nations nor even, until the late fifties, the Council of Europe (created in 1949). In spite of this refusal to join the mainstream of contemporary internationalism her foreign policy, while not popular, was considered to be at least consistent and her defense policy at least credible. Swiss neutrality gained in esteem when, in 1955, the great powers agreed at long last that Austria could opt for a neutrality shaped after that of Switzerland.

The attitude of the Soviet Union toward neutrality and the neutrals has undergone various phases and changes as well. Unlike that of the Western powers the Soviet attitude was, however, much more conditioned and shaped by ideological concepts. Here, as in other cases, theory had to adjust to reality. Both in the legal and the political field the Soviet Union went some way to modulate and differentiate her views on neutrality and, incidentally, also on nonalignment. It is therefore misleading to describe Soviet policy toward the neutrals as immutable: it is not. What has not changed so far is Moscow's fundamental rejection of neutrality (and nonalignment) for home consumption, i.e., neutrality as a policy option for any of her present allies in Eastern Europe. Nor has the conviction changed that outside the Soviet orbit neutrality can, if properly handled, be put to use for Soviet interests. It may be marginal for central strategic objectives but useful on the operative and tactical level.

Neutrality and its future will therefore remain on the political agenda of the Soviet Union in various guises and contexts. It is one of the purposes of the present chapter to examine whether and how Soviet policy under Gorbachev will take a new look at the place and role of the neutrals in a changing European environment and to what extent this could affect the neutrals' outlooks and policies.

NEUTRALITY VERSUS NONALIGNMENT

Neutrality in Europe, it was said, takes several forms. And so does nonalignment. Simplifying somewhat, one could say that neutrality has its basis in international law and deep roots in European history; nonalignment has its origins in international politics as it was shaped by the East-West conflict and the process of decolonization since 1945. Swiss and Swedish neutrality are perhaps the best examples of the former; Yugoslavia, as one of the leaders of the nonalignment movement, is, certainly in Europe, the most prominent example of the latter.

Neutrality, because of its legal status and its tradition, is no doubt more predictable and stable than nonalignment. That is the reason why the Western powers have generally preferred it; and that is why the Soviet Union has been more inclined or tempted to support nonalignment, at least in the developing world. After having discovered and accepted this part of the world as a new factor of international politics, the Soviet Union also recognized the advantage of using the political fluidity of nonalignment for her own purposes. The wish of the nonaligned countries to stay out of the East-West conflict was interpreted by Moscow primarily as their refusal to accept Western influence in general and any association with Western political or military alliances in particular. Given the Western—above all, the American—dominant presence in the Third World, such an attitude was in itself a welcome assurance for the Soviet Union.

But above and beyond such a policy of denying influence to the West, there was a more fundamental, though often ignored, consideration behind Soviet predilection for nonalignment. It has been pointed out that the real difference between the neutral and the nonaligned states is not so much one of legal status or of foreign policy[4]—though obviously these differences matter as well. Rather, it lies in the way these two groups of countries perceive the international order and its future evolution. The former—i.e., the European neutrals in the first place—basically accept the existing order and feel as an integral part of it. Finding themselves embedded in, and in various degrees also protected by, the existing security system on the continent, they have a stake in maintaining and improving the status quo rather than upsetting it. Their active participation in the Conference on Security and Cooperation in Europe (CSCE) and their signing of the Helsinki Final Act in 1975 are perhaps the clearest manifestations of this attitude.

The neutrals' position contrasts in many ways with that of the nonaligned countries. Hardly any of the latter feels responsible for the international order into which they were born. Many thus reject it altogether or want to see it changed in favor of what they hope to be a more equitable and just order. In a sense, then, they are against the present status quo. In this they find an ally in the Soviet Union, and vice versa. In the heyday of the nonalignment movement it mattered little where this common travel would lead to as long as both parties agreed on who had to pay the price for it, i.e., the West. But disenchantment on both sides was bound to grow when the Third World realized that the Soviet Union either did not deliver or joined hands with the West, above all the United States, when it suited her own interests (as in the case of the nonproliferation treaty). The Soviet Union in turn discovered that many of the developing countries were not prepared to travel the road toward socialism as she would have liked them to do. They began to pursue, however imperfectly, a policy of nonalignment not only vis-à-vis the West but vis-à-vis the Soviet Union as well.

Against this background, Soviet policy toward the European neutrals and its variations become more intelligible. While never fully abandoning the idea that neutrality in Europe, too, was open to change and influence, the Soviet Union entertains less hopes with regard to the scope and velocity of such change and the degree of influence she may exert to bring it about. In general, she became realistic enough to consider neutrality, as practiced by the European neutrals, something less volatile than nonalignment. To be sure, she would not hesitate to play the card of neutrality should she consider it useful in some instances and periods of East-West relations. But as part of a well-established strategic situation, Moscow accepts that neutrality allows only for marginal shifts and practically no interference. While it is premature to say how Soviet policy under the new leadership will be conducted in the Third World and toward nonalignment, it can be said that it has become noticeably more pragmatic toward Europe's neutrals.

A NEW SOVIET PRAGMATISM

Soviet pragmatism is neither unconditional nor undifferentiated. The Soviet Union is well aware of the considerable nuances in the legal foundation and political practice of the four European neutrals (or five, if Ireland is included). And she acts accordingly. Thus, she recognizes—as mentioned before—the linkage between Finland's neutrality and that of Sweden on the one hand, and that of Switzerland and Austria on the other.

The first pair became part of what later (and somewhat euphemistically) was called the Nordic balance. It describes a strategic relationship in which the North Atlantic alliance appears as a fairly discrete partner in the West, and the Soviet Union as a direct participant in the East. Having once accepted the special brand of Finnish neutrality, the Soviet Union would have been ill advised not to respect that of Sweden as well. In different ways both turned out to be an advantage for her: Finland as an example and showcase of Soviet tolerance and acceptance of a neutral neighbor; Sweden because any other status than neutrality could, given her geographic location and her considerable economic and military weight, change the Nordic balance most likely against Soviet interests. This is particularly true should Sweden decide to join the Western alliance. If that happened, it is hardly conceivable that Moscow would not review its attitude toward Finnish neutrality. In turn, it is equally difficult to imagine that Sweden would remain indifferent should the Soviet Union draw Finland into her orbit. Thus, realistic options for a change of political course on either side are limited indeed. The Soviet Union is certainly aware of this; but it has not condemned her to immobility, as will be shown.

The combination of the "alpine neutrals," Switzerland and Austria, is of a different kind. It is, as far as we can make out, also differently perceived in the Soviet capital. Together the two form a neutral corridor in the heart of Europe that finds its nonaligned prolongation in Yugoslavia. They split the alliance into two parts, thus making communications more difficult for the West and dividing the front in the East. But unlike the neutrals in the north, the strategic utility of Austria and Switzerland is minor for the Soviet Union; she does not have the same leverage on them as she has toward the Nordic countries. In contrast to the Nordic balance, of which she is a part, the central European neutrals are beyond her influence though obviously there are important differences in the positions of Switzerland and Austria. Few, if any, alternatives to the existing situation offer themselves to the Soviet Union: her room for action is limited in times of peace.

Switzerland's neutrality is well entrenched in European history and politics. It owes its duration and credibility to an almost unique convergence of interests between the foreign powers and Switzerland, the former wanting to keep Switzerland and her mountain passes out of mutual competition, the latter seeing in her neutrality the only means to maintain internal cohesion and avoid external entanglement. Both have good reason to believe that the extent of the Swiss defense effort raises the price of aggression to a level that is disproportionate to the gains that can be expected from it.

The Soviet Union would also be hard put to justify any attempt to question Austria's neutrality, having herself suggested that it be modeled after the Swiss example. Austria, not unlike Finland, is used by the Soviet Union as a latter-day example of her recognition of neutrality as an acceptable political option for certain European countries. While she may reverse this decision under changed conditions, she has little reason and even less excuse to do so in a situation of overall European stability.

If in geographical terms we find some coupling between the two north European neutrals on the one hand, and the central European ones on the other, there is, in terms of status, a different kind of affinity between Finland and Austria as against Sweden and Switzerland. The Soviet Union sees the neutrality of the latter pair as less vulnerable and possibly more resilient to outside influence. With regard to Austria and Finland, Soviet policy is primarily one of denial: to prevent one or the other from doing what runs counter to Soviet interests. This would be the case if either country acquired weapons which, by treaty obligation, they are not entitled to; it may be also the case if they wanted to become full members of the European Community at the price of abandoning their neutrality. How strong and enduring Soviet resistance to such moves will be is a matter of conjecture. It will depend on both the nature and direction of Soviet policy and on the changes that Europe may be facing in the years to come.

A CHANGING EUROPEAN ENVIRONMENT

Indeed, the political climate in Europe in recent years has undergone tremendous change. And more seems to lie ahead. We may be witnessing what I would call, for lack of a better word, a restructuring of postwar Europe. There are many reasons that make such a prospect likely. We see a Soviet Union in the throes of a potentially far-reaching social, economic, and political transformation (perestroika). We see a United States, weakened by budget and trade deficits, reappraising its foreign commitments. As a consequence, both powers are reviewing their policies toward Europe: the Soviet Union because she wants strategic stability in exchange for economic cooperation and Western technology; the United States because it feels that Western Europe has become more of an economic competitor and less of a financial contributor to common defense. Both seem to have recognized that the excessive size of their nuclear arsenals is now out of proportion with their political utility. Their agreement on the total elimination of medium- and shorter-range missiles (the INF agreement) is a first step; others may follow. It has set in motion not only a reappraisal of the dangers and costs of an unending arms race but, more interestingly, a willingness, on both sides, to consider for the first time the possibility of conventional disarmament in Europe.

If the first challenge to Europe therefore comes from the changing moods and interests of the superpowers, the second comes from the prospect of disarmament, both nuclear and conventional. The third challenge is likely to come from the creation of a single market among the twelve members of the European Community, scheduled to be accomplished by 1992. Whether or not this deadline will be met, the political will to do so is beyond doubt. Such a single market with its unhindered flow of people, goods, services, and capital is bound to change not only the economic but also the political landscape of Western Europe and, most likely, beyond.

To be sure, none of these developments, important as they may be, will do away with the division of Europe into two politically and ideologically different camps. In this sense, if in no other, Europe will remain divided and the antagonism between East and West will endure. This basic reservation apart, we can expect the Europe we have known for more than forty years to change in many respects. How fast and in what way this will happen, nobody knows. Nor do we know what the new Europe will look like. Perhaps it will have a reduced presence of the United States and a more open and less threatening Soviet Union; perhaps there will be more political and defense cooperation in Western, and more instability or unrest in Eastern, Europe; perhaps there will be less, but more modern, nuclear and conventional forces and new forms of East-West relations.

If these are going to be the contours and directions of change in Europe, they must interest the neutrals as well. These countries will have to ask themselves whether or not they should join the European Community, whether and when they should take part in negotiations on conventional disarmament, and whether changes in the policies of the superpowers will affect their security. They may give different answers to such different questions, but all are faced with one and the same challenge: to determine whether, in such dramatically changing circumstances, they can preserve their neutrality or must modify and possibly even sacrifice it.

In a sense Europe—and with it the neutrals—is faced with two different visions of the continent's future political structure: on the one hand we find the objective of the creation of a single European market, on the other stands Secretary General Gorbachev's idea of a "common European house." The two concepts not only differ in many ways; they are, in final analysis, diametrically opposed.

The moving forces behind the creation of a single market are the promise of further economic progress and a more evenly spread prosperity, the abolition of outdated barriers of all kinds, and the mobilization of yet untapped common energies. This effort at a more and more comprehensive integration should lead, in the eyes of many Europeans, eventually to a political unification of Western Europe or at least of those countries which will then be members of the European Community. It is admittedly a distant goal, unlikely to be reached before the turn of the century. But it would be the logical consequence of a long and often delayed process initiated with so much enthusiasm immediately after the war.

However tempting such a goal may be, its price could be heavy. It could deepen rather than bridge the East-West divide, leaving Eastern Europe possibly forever at the mercy of Soviet dominance and removing any prospect of German reunification.

Gorbachev's vision of a common European house is at the same time much more vague and ambitious than the aim of West European integration. It envisions a Europe that should adopt a common system of security buttressed by an ever-growing mutual interdependence among the European states. But Gobachev and his associates have so far stopped short of giving any details about how this house should be constructed and maintained, who its tenants are, and how and by whom it should be run. In other words, the vision is there but we know little about its substance.

Gorbachev's European house must interest the neutrals more than anyone else. Nothing is said about the place and role they will be accorded in the house or about the conditions for their admission. Until further notice we must therefore assume that the Soviet leadership tacitly accepts the neutrals as a given group of states that forms a part of Europe and will do so for any foreseeable future. This suggests that the Soviet Union, until further notice, does not anticipate any fundamental change in the neutral's political status, nor does she herself seek such a change.

COMING TESTS FOR NEUTRALITY

It is these two concepts that the neutrals face today and probably even more so tomorrow. They may temporize on the former, i.e., the single market, and not take seriously the latter, i.e., Gorbachev's common house. Whether this is a wise policy remains to be seen. But the neutrals cannot afford to ignore one or the other. Europe with a single market and a politically strengthened European Community—be it in 1992, as envisaged, or a few years later, as seems likely—will be a vastly different place from the Europe the neutrals have known and learned to live with since 1945. If they stay out, they risk isolation or marginalization; if they join, they risk a diminution or revision of their neutral status.

What interests us here is not so much the choice the neutrals will make themselves but the reaction of the Soviet Union. None of the neutrals can act totally on its own when it comes to joining the European Community: whatever one does will affect in one way or another the position of the other neutrals.

There are differences: Sweden and Switzerland have, no doubt, a greater margin to maneuver than Finland and Austria, whose neutrality, for reasons of their special relations with Moscow, is more narrowly circumscribed. As mentioned before, the Soviet Union, more than any other state, is watching carefully every movement of these two countries when it comes to their policy of neutrality. It is far from certain that she would acquiesce to their entry into the Common Market without protest. While having become more pragmatic in her attitude toward the European Community—witness the recent agreement between the community and the Council for Mutual Economic Assistance (COMECON)—she assumes, probably rightly, that it will sooner or later expand into the field of defense and foreign policy. In this case any neutral country who joined the community would see its neutrality, however well preserved, put in jeopardy if not discredited.

It is difficult to visualize the Soviet Union accepting such a basic change of postwar Europe's political configuration without any attempt to prevent or oppose it. More than anything else, such a prospect would highlight the role she has played in preventing Europe from being simply a juxtaposition of two politically and ideologically antagonistic camps, thus contributing to its stability and diversity.

The merging of Europe's neutrals with a politically ever more coherent European Community and, as a result, the disappearance of neutrality, would be seen as such a major change. How the Soviet Union would react to such an event is difficult to predict. It largely depends on the circumstances in which the neutrals (above all Austria and Finland) would take such a step and on the overall political climate in which they did so. With the prospect of disarmament and greater security on both sides, the Soviet Union, too, might feel less worried about such a change than she would in a climate of renewed external hostility and sharpened internal unrest. Which of the two alternatives is more likely in the years to come is too early to tell.

But even if the neutrals stay out of the Common Market, or succeed in preserving their neutrality while joining it, their neutrality will be put to a severe test. This is not to say that neutrality has outlived itself. But it will have to prove that it can still make a contribution to European stability and the promotion of the East-West dialogue rather than become more of a burden to one or both of them. So far, it would seem that the neutrals still have a role to play and that this is accepted by all parties concerned. How long this will remain so depends as much on the changing environment as on the neutrals themselves.

But neutrality is, after all, not an end in itself. Rather, it is the means that countries like Sweden and Switzerland, Austria and Finland have chosen in order to maintain the greatest possible degree of national dependence and prosperity. If either of them is threatened because neutrality no longer serves them, then the time has come to look for those means which can do it better.

All this is looking far ahead. But the fact that Austria is seriously considering the possibility of joining the European Community shows that things are beginning to move. This may not signal the end of neutrality: the Austrians themselves want to preserve it even as member of the community. But the neutrals will have to take a hard look, either individually or jointly, at the challenges common to all of them, and demonstrate that in the midst of a changing environment their neutrality is still the best option—for them no less than for Europe as a whole. If they succeed, the chances are that both the West and the Soviet Union and her allies will accept it. They, too, may feel that the "missing link," the neutrals, still have an important function to play and that Europe would be the poorer if they prematurely ceased to play it.

CHAPTER EIGHT

EUROPEAN NEUTRALS IN SOVIET MILITARY STRATEGY

John G. Hines and Phillip A. Petersen

Since Mikhail Gorbachev became general secretary of the Communist party of the Soviet Union (CPSU) in 1985, the traditional zero-sum security assumptions that have dominated the Soviet view of international relations increasingly have been challenged by highly placed advocates of a broader approach to questions of national security. Proponents of this new thinking have given greater weight to political, as distinct from purely military-technical, variables in the security calculus and support a policy that emphasizes threat reduction, unilateral restraint, and collaboration with adversaries.[1] Even the military, which consistently resists any "unilateral activity to establish equitable defense sufficiency," acknowledges that "assessing security is more and more becoming a political task" to be "resolved by political means through detente, disarmament, strengthening confidence, and developing international cooperation."[2]

It is possible that Gorbachev and his associates will succeed in permanently altering the relative weight given to the military and political components of security in Soviet military doctrine. They are very likely to succeed in lowering NATO and Warsaw Pact force levels in Europe even if their doctrine is not substantively altered. Under these circumstances the relative operational significance of the contribution by neutral states to strategic stability will increase. In Soviet thinking, in the event of war a military doctrine of sufficiency and defense would give way immediately to a strategy of offensive or counteroffensive operations. In this context the potential impact of the neutrals on Soviet ability to execute the operational concepts in its military strategy becomes the ultimate measure of their contribution to deterrence. Moreover, it is possible to evaluate the role of these states in a Soviet offensive strategy since even Soviet analysts now argue that it is inconceivable that NATO would attack the Soviet Union.

To evaluate the contribution of the neutral states to crisis stability through deterrence, this chapter examines European defenses and terrain[3] within the framework of possible Warsaw Pact attack variants that are consistent with our understanding of the Soviet approach to offensive operations. Because geography dictates that the neutral states would probably be the first to be violated by initial WP offensive operations, our analysis focuses on the first phase of WP operations in the various theaters of strategic military action (TSMA) (in Russian *teatr voyennykh deystviy* or TVD) into which Soviet planners divide Europe.

THE STRATEGIC OPERATION IN THE NORTHWESTERN TSMA

Consistent with the Soviet General Staff view of the nature and importance of a strategic region, northern Europe can be said to contain four strategic regions: (1) northern Norway, Sweden, and Finland; (2) southern Norway, Sweden, and Finland; (3) the Baltic Straits; and (4) the Greenland-Iceland-United Kingdom-Norway Gap (see fig. 8.1). These four strategic regions draw together the operational objectives of four Soviet theaters of strategic military action: Northwestern, Western, Arctic, and Atlantic.

Most Western analysts seem to agree that it is unlikely that northern Europe would become embroiled in a war separate from a war in the Western TSMA.[4] It is very likely that a war in the Western

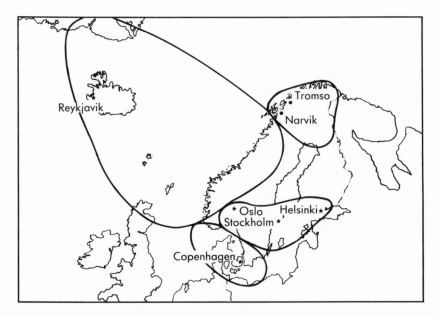

Figure 8.1 Strategic Regions in Northern Europe

TSMA would escalate horizontally into the Arctic Ocean TSMA.[5] Such combat actions in the Arctic Ocean TSMA would undoubtedly include air strikes against its "contiguous coastlines" along Norway as well as against Iceland as an island in this oceanic TSMA.

Soviet combat actions in the Arctic Ocean TSMA would relate to three strategic missions. The first is to support and protect the ballistic missile submarine (SSBN) force, and thus insure its viability as a survivable retaliatory component of Soviet strategic nuclear forces. The second is to defend the homeland from attack. The third is to contribute to the multitheater wartime objectives of the Soviet Union, focusing on operations in the Western TSMA. Many of the wartime tasks required by any one of these missions would serve the other two. One example is anticarrier warfare. The United States carrier battle groups threaten the Soviet SSBN force by giving support to U.S. antisubmarine warfare operations. The nuclear and conventional strike aircraft onboard the carrier threaten both the Soviet homeland as well as Soviet forces operating in the Western TSMA. By successfully attacking the carrier, all three missions are served.

The mission of supporting and protecting the SSBN force is important no matter what the nature of a war might be. The Soviets anticipate that even during conventional war both sides would be striving to destroy the other's nuclear capabilities with various non-nuclear means.[6] Under such conditions, reliance on the SSBN force as the core of a survivable strategic nuclear capability is threatened by the U.S. advantage in antisubmarine warfare and, during any conventional war lasting several months, such vulnerability would be greatly increased when the submarines attempted to replenish. The deployment of the SS-25 mobile ICBM may provide at least a partial solution to the Soviet problem. Land-mobile missiles operating from presurveyed positions potentially could have accuracy approximating that of silo-based ICBMs and be far more survivable over time than Soviet SLBMs.

Northern Norway, Sweden, and Finland

If the Soviets do come to perceive it to be in their interest to horizontally escalate a conflict into the Northwestern TSMA, an initial offensive early in a conflict would undoubtedly focus upon northern Norway (see fig. 8.2). This initial strategic direction by the Arctic front would probably consist of a southern operational direction and a northern operational direction. In the south the Soviets would probably only engage in actual combat action if necessary to prevent Finnish military forces from reinforcing the north. In the north the Soviets would attempt to move on Finnmark as quickly as possible through the use of forward detachments, air assault units, and airborne forces. Although the Soviets would prefer to pass through Finnish territory without engaging its armed forces, the Soviets appear to accept such a possibility as a necessary risk.

The operational direction in the north can be divided into four main axes of advance (see fig. 8.3). Moving from east to west, the first axis of advance would be on Kirkenes and along the coastal highway toward Lakselv. Movement would have to be confined to the coastal highway in many places, but the country would not be

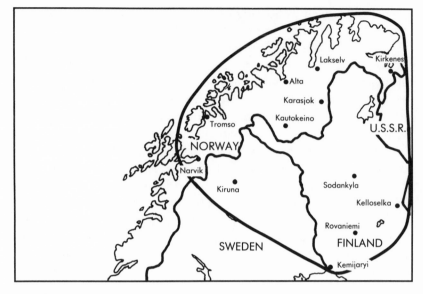

Figure 8.2 Northern Strategic Region in Northwestern TSMA

very different from that of the Petsamo area of the Soviet Union close to the border. Special units such as heliborne and amphibious forces could be used to effect early seizure of critical objectives to facilitate movement.

All other axes of advance would cross the Soviet border south of Lake Inari. A significant proportion of this terrain, estimated by the Finns to be as much as 50 to 60 percent of the area,[7] is swampland covered with numerous streams and lakes. In winter the ground freezes before being insulated by snow, on the average, only once in seven years. Thus, crossterrain transit is made even more treacherous by winter conditions hiding soft ground. Even lightly armored reconnaissance vehicles would have difficulty negotiating this terrain at any time of the year.

The easternmost axis of advance toward Karosjok crosses the Norwegian border to two all-weather roads leading up a broad valley. At Karosjok the axis splits, with one direction moving north toward Lakselv. As this axis of advance nears Lakselv, however, the terrain turns more difficult. While the coast near Lakselv would be conducive to the conduct of amphibious envelopment, the terrain to the south of Lakselv still seems to make the area a reasonably good defensive position. The other axis moves east from Karosjok to meet the north-south road between Alta and Kautokeino, before turning north to Alta. As this axis of advance moves down off the high broad plateau, called Finnmarksvidda, toward Alta and its airfield, the terrain gets very difficult, with the main road flanked by a river and sheer cliffs.

Even if the Soviets should succeed in an attempt to seize Lakselv and Alta, they will have gained little except longer supply lines. Furthermore, the loss of the airfields probably would not have any great impact on NATO's prosecution of the war at sea. The only real issue would be the price the Norwegians had been able to extract from the Soviets. To stiffen the resolve of the Finnish and Swedish politicians, the Norwegians need to show that the Soviets might not

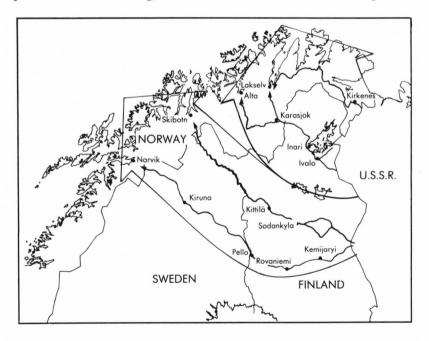

Figure 8.3 Probable Main Routes of Advance in Northern Strategic Region of Northwestern TSMA

win. An unopposed occupation or easy victory at Lakselv and Alta could strengthen the voices of accommodation in Finland and Sweden. The Finns and Swedes would need to know the sacrifices that would be asked of them by their Nordic neighbor would make a difference in the long run. If the Soviets were to encounter difficulties taking Lakselv and Alta, it would not bode well for them, given the defenses they could expect to encounter farther west.

The remaining two axes of advance against northern Norway must pass through the heart of Finnish defenses in Lapland. The third axis would initially encounter low terrain with numerous water barriers and, probably, defensive lines at Sodankyla and Kittila. Ultimately, transiting the mountainous terrain of western Finnish Lapland, this third axis of advance would be directed toward Tromso through the Skibotn valley, which is very difficult terrain. There are routes south of the valley that are easier initially, but these subsequently become substantially more difficult.

A fourth axis of advance on this northern operational direction in the Northwestern TSMA might be directed through Kiruna, Sweden, toward Narvik, Norway. This axis would face major water barriers at Kemijarvi and Rovaniemi, Finland, and lakes also dominate the axis from Rovaniemi to Sweden. As in much of Finnish Lapland, off-road travel in northern Sweden is often poor. The roads to Kiruna cross numerous water barriers, and the Swedes have prepared their bridges and roads for demolition. Such demolition could not only slow a transiting enemy's rate of advance, but also expose his logistical tail to devastating disruptions because of the often swampy character of the off-road terrain. Although an all-weather road has recently been constructed between Kiruna and Narvik, the terrain on this axis of advance remains difficult.

Degraded reliability of combat vehicles, crew exhaustion, and shortage of logistics would have already left Soviet forces ill prepared for a frontal assault against Fortress Norway (Troms area). Clearly, Soviet assault landings in some form would be necessary to cut off prepared Norwegian defenses from reinforcement.

The Norwegians have invested heavily in coastal fortifications which, together with their navy and air force, would make any naval assault landing difficult at best. A small special purpose (*spetznaz*) force might be delivered through the north-central Norwegian archipelago to the mainland by submarine, which would be hard to detect in coastal waters. Any large-scale assault force, however, first would have to avoid detection by the air force and navy, and then would have to fight its way past a coastal fort. Once engaged and fixed by the fort's long-range weapons, the amphibious assault force could be attacked by Norwegian air and naval forces. Even if Norwegian airfields would have been suppressed, Norwegian navy fast patrol boats could hide in the archipelago and would be difficult for Soviet attackers to detect and engage with missiles before being brought under attack by the Norwegian boats' on-board guns and missiles.

Vertical envelopment offers another possibility for adding weight to the Soviet attack on Fortress Norway. Obviously, if the West were to have local air superiority, the Soviets would have great difficulty with insertion of a force large enough to make any substantial contribution to the offensive. If the West should fail to attain air superiority, however, deficiencies in Nordic ground-based air defenses could allow the Soviets to use air mobile operations to out maneuver the defenders. The Bofors RBS-90, if deployed widely enough in Finland, Sweden, and Norway may, however, to a large extent resolve this contemporary vulnerability.

Much of Finnmark consists of wide-open conditions in which cover is sparse. Concealment of static forces is possible, but moving formations are easily spotted. Any Soviet attempt to seize northern Norway thus depends on a nonhostile air environment.

Provisions of the Paris Peace Treaty of 1947 restrict the Finnish air force to only sixty combat aircraft. Because only about one-third of this force (the Lappi Air Wing) is allocated to protect the northern part of the country,[8] the Soviets might perceive that they would not be taking that great a risk by not striking Finnish airfields with the initiation of hostilities. If the Soviets find that they must suppress Finnish air activity in the north, such a requirement might be limited to the operational zone of the Lappi Air Wing. Aircraft in the Finnish northern air defense area are sheltered and widened parts of highways have been prepared for use as backup runways.

Sweden has a well-trained and modern air force of some five hundred combat aircraft.[9] It is probable, therefore, that the Soviets would go to great lengths to avoid drawing Sweden into any war, at least early in the conflict. There is a penalty, however, for not attacking the Swedish air force preemptively. Once alerted in a crisis, Swedish aircraft could be dispersed to numerous wartime bases and present the Soviets with a formidable targeting problem. Many of these bases have been integrated into the highway system to enhance aircraft dispersal and to provide a larger number of runways that would have to be neutralized in order to shut down operations at a single base. At some of the bases the Swedes even have constructed underground aircraft shelters.[10]

NATO reinforcement would also be a pressing question for Soviet planners seeking to neutralize the air threat to their forces operating in the far north. The Soviets note that Norway has far greater airfield capacity than required by the Norwegian air force, enabling large numbers of aircraft to be relocated to the region in the event of crisis. They appear to be convinced that, as a minimum, Soviet forces must be prepared to neutralize the airfields at Banak, Bardufoss, and Boda.[11] This task has been made significantly more difficult by the underground shelters and aircraft bunkers constructed at Norwegian airfields. As a result, mining of the airfields would appear to be the most effective near-term Soviet solution to the requirement to suppress aviation activities at Norwegian bases.

All concerned seem to recognize that the Northwestern TSMA is not where general war between the two main military blocs would be decided. Each side might hope, nonetheless, that its actions might cause the opposing coalition to divert sufficient resources to the Northwestern TSMA to greatly decrease its chances for victory in central Europe. If NATO could be persuaded that its scarce reserves must be deployed in north-central Norway, they would not be available for the battle for Denmark. Thus, Soviet planners would probably take into account that a favorable change in the correlation of forces in the Western TSMA might be precipitated by their initiating action in the Northwestern TSMA. On the other hand, the more Soviet resources are drawn away from central Europe, the greater the chance that the Soviets will be denied a quick victory at the conventional level.

The Swedes require special consideration by Soviet planners. If

Sweden's stated policy of neutrality becomes one of armed opposition, it could significantly alter the correlation of forces in the Northwestern TSMA as well as in the Baltic Straits strategic region of the Western TSMA. Thus, Soviet action against northern Norway across Sweden could have a severe negative effect on operations in the Western TSMA.

Southern Norway, Sweden, and Finland

The most important administrative-political and industrial-economic centers in the Northwestern TSMA are in the southern region of the Northwestern TSMA. In addition, airfields in southern Norway and southern Sweden could make a considerable contribution to the defense of Denmark. As a result, Soviet combat actions against this strategic region would probably be tied to the scenario in the Western TSMA. Soviet combat actions against the region could be directed north to south, down the Scandinavian peninsula. The Soviets might also strike against southern Norway by means of a joint amphibious and airborne assault landing operation initially directed against southern Sweden. Apparently, Soviet World War II General Staff studies anticipated the possibility of such a maritime direction.[12] Soviet forces executing such combat actions today probably would be isolated logistically if they were not immediately successful. The long range of modern Western combat aircraft would enable opposing air forces to operate against assault landing forces out of Norwegian airfields if necessary.

The Baltic Straits

In Soviet planning the Baltic Straits could be a strategic region of either the Northwestern or the Western TSMA. In the Soviet view, however, seizure of the straits would normally be a strategic objective in the Western TSMA. Any Soviet assault landing operation against Zeeland would be very vulnerable to air strikes flown from airfields in southern Norway and Sweden. Neutralizing these airfields would prove very costly in aircraft if air strikes were conducted up the

Kattegat and across the Skagerrak Straits against Norwegian air-fields.

Air attacks across southern Sweden could prove equally costly and would probably require the Soviets to make at least limited air strikes against Swedish airfields. Even a single successful mass strike that succeeded in dropping mines on the airfields of southern Scandinavia, however, might provide suppression of Norwegian and Swedish airfields sufficient to allow the Warsaw Pact to land assault forces on Zeeland with enough air defense capability to greatly repulse subsequent air attacks. As noted earlier, however, the costs of bringing Sweden into the war as a cobelligerent with NATO could be very high in the long term.

Success in taking Denmark out of the war would not necessarily insure the Warsaw Pact naval forces passage in or out of the Baltic Sea. Airfields in the United Kingdom, Norway, and, if it becomes a cobelligerent, Sweden would still have to be suppressed or Warsaw Pact ships attempting to enter or exit would have to run a formidable gauntlet. The seizure of the Baltic Straits, however, would insure that NATO could not use Danish territory and airspace to strike deep into the Warsaw Pact operational flank in the Western TSMA. The gains could be offset, however, if in the process of achieving them the Swedes were pushed into joining Finland as a cobelligerent with NATO against the Soviets.

THE STRATEGIC OPERATION IN THE
SOUTHWESTERN TSMA

According to a Soviet military academy lecture presented in the mid-1980s, "the operational directions of the Southwestern TSMA are determined on the basis of the peculiarities of the natural conditions, the condition of the road network, the locations of operational-strategic objectives, experience of the Great Patriotic War, and the present political situation." Possible operational directions are said to include the following:

1. Alpine direction: frontage—180 to 240 kilometers; depth—550 kilometers; operational density—six to eight divisions.

2. North Italian direction: frontage—100 to 200 kilometers; depth—600 to 750 kilometers; operational density—six to eight divisions.

Alpine Operational Direction

The Alpine axis is directed at the task of defeating and occupying Austria. The execution of combat action on this operational direction is, in all probability, tied to the willingness of the Austrian government to tolerate the transit of Warsaw Pact forces through its territory on their way from Czechoslovakia to Germany and from Hungary to Italy. If the Austrians should insist upon defending their neutrality, the forces on this operational direction would endeavor to force inaction or to neutralize Austrian forces.

By crossing the Danube River in Czechoslovakia, Warsaw Pact forces could move against Vienna from both sides of the river. The Vienna airport lies to the southeast of the city, and the city itself is bordered by elevated terrain to the west and south. As a result, the city and airport probably could not be defended successfully.

If the Austrian government were not captured and the Austrian armed forces were to continue to resist, Warsaw Pact forces would probably move up the Danube River valley to link up with forces fighting in the Western TSMA. Subsequent combat actions by the Warsaw Pact would be directed at ending Austrian resistance. If this resistance proved stubborn, the Warsaw Pact would probably opt for a temporary strategy of controlling key lines of communication needed to support operations being conducted against Germany and Italy.

The Austrians appreciate that "future aggressors will be interested primarily in controlling the main communication lines" and, "therefore, subsequent zones along these lines have been prepared for static defense by fortifications and prepared demolition to achieve both delay and attrition of the aggressor."[13] However, deficiencies in deploying modern technologies severely threaten the credibility of Austria's capability to execute its strategic concept.

North Italian Direction

Operating to the immediate south of the Alpine direction is the north Italian operational direction. If at some point in a European conflict Soviet leaders concluded that it would be necessary to occupy a part or all of Italy, such combat actions would be initiated over routes through Austria and Yugoslavia to northern Italy. This axis would take the Warsaw Pact forces through Italy to link up with WP forces of the Western TSMA advancing into southern France.

THE STRATEGIC OPERATION IN
THE WESTERN TSMA

The Soviets assess the Western TSMA to be the main theater in a war between the Warsaw Pact and NATO. They seem to believe that NATO is a strong military alliance, but a weak political coalition and some aspects of their operational planning reflect an attempt to exploit perceived NATO political vulnerabilities. The Soviets understand that major NATO defense decisions require agreement of the allies and would hope to aggravate the confusion and anxiety that would likely characterize NATO responses in the opening period of war in Europe. They would strive to introduce delay into the NATO decision-making process sufficient to allow the Warsaw Pact to obtain its immediate objectives before effective defensive action could be taken.

For political and geographical reasons, NATO contingents (organized in national corps) each holds a "slice of the line" on the central front. These national corps sectors lie along the intra-German border (IGB) and along the border between the Federal Republic of Germany (FRG) and Czechoslovakia. This strategy of forward defense is a reflection of the political requirement to convince the German public that it is in fact defended. Forward defense is also an operational imperative, however, in that the terrain near the IGB and FRG-Czech border is more defensible than that terrain located at greater depth. Mountainous and hilly regions and extensive wooded areas along the borders restrict the amount of force the Warsaw Pact can bring to bear in any one place. Farther west, the terrain opens up to present an attacker the opportunity to bring greater force to bear upon the defending forces.

A Warsaw Pact strategic offensive in the Western TSMA would consist of several joint and combined arms operations performed in accordance with a single concept for the defeat of the NATO coalition in the theater. A strategic operation in the Western TSMA would include the following types of component operations: air, antiair, front, assault landing, and naval. If NATO attempts to use nuclear weapons, the Warsaw Pact strategic operation may also include nuclear strikes.[14]

An air operation (*vozdushnaya operatisiya*) is a non-nuclear operation to destroy or weaken the enemy air and nuclear forces. In order to do so, it also would have to neutralize the enemy's air defenses. It would include not only aviation strikes, but also strikes by artillery and missiles, as well as assaults by airborne, heliborne, and *spetsnaz* troops.[15]

From the Soviet perception, NATO defense in the Western TSMA presents them with what is essentially a tactical problem. In wartime, the bulk of NATO's ground forces, and about 50 percent of NATO's firepower would be deployed in what the Soviets refer to as NATO's tactical zone of defense (the first 30 to 50 kilometers). Obviously, it would be the Soviet aim to place the main weight of attack on the weaker areas of this shallow, sectored line of defense.

By calculating coefficients for the quality of the various NATO divisions (see fig. 8.4), the Soviets identify those corps sectors that should require the least effort to penetrate. In this way the Soviets would attempt to identify for attack the enemy's less capable forces and would engage the enemy coalition's stronger forces with active defensive operations or limited attacks. The Soviets typically would strive to encircle the bypassed stronger forces to attack them from the flanks and rear, prevent their escape, and ensure their destruction. Thus, the application of basic Soviet military planning principles to possible Warsaw Pact offensive operations in the Western TSMA suggests that some WP attack variants are more likely than others. Specifically, the Soviets probably would attempt to penetrate the Dutch, Belgian, and British corps sectors in the hope of encircling the U.S. and German corps.

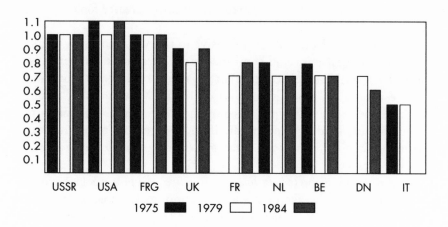

Figure 8.4 Soviet Division Coefficients for Measuring the Correlation of Forces: 1975, 1979, 1984

According to lecture materials from the Soviet General Staff academy, the Western TSMA may be envisioned as being made up of two strategic directions: north and south Germany. The operational capacity of each of these strategic directions is said to be "sufficient for the deployment and military operations of up to two *fronts*."

North German Strategic Direction

The north German strategic direction is probably directed against NATO's Northern Army Group (NORTHAG) and those Allied Forces North (AFNORTH) contingents located in Denmark and the federal German state Schleswig-Holstein. These latter forces come under the operational control of the NATO Commander Baltic Approaches (COMBALTAP), who is a Danish officer. NORTHAG comprises four corps, each of which is provided by a different NATO member nation. The German corps, assessed by the Soviets to be the strongest, is in the center and those assessed to be less combat capable, the Dutch and Belgian, are on the NORTHAG flanks. Any Warsaw Pact attack against NORTHAG, therefore, could be expected to attempt to encircle the German corps.

The northern or coastal front on the north German strategic direction would probably have two initial operational directions: one directed against Schleswig-Holstein in AFNORTH and a second directed at the Dutch corps sector in NORTHAG. The operational axis directed against Schleswig-Holstein would be aimed at seizing the probable immediate army objective, the Kiel Canal. While numerous tactical obstacles exist, the terrain in general would pose no particular difficulty. For the most part, the terrain on this axis is flat. Tactical scale amphibious assault combat actions would undoubtedly be employed to assist the ground advance on this operational direction.

The initial key terrain features to be considered in the Dutch corps sector are the Elbe River, constituting the northern boundary, and the Elbe-Seiten Canal, running north and south across the entire corps sector (see fig. 8.5). The Soviets probably would plan to cross the Elbe River on both the Pact and NATO sides of the IGB. By attacking the Dutch corps from the south, the Warsaw Pact could exploit to some extent the boundary between the Dutch corps and the I German corps. Such an attack should fix the attention of the Dutch corps on the defense of Uelzen. At this point the Dutch would be able to make use of the combination of the Elbe-Seiten Canal and the north-south rail line to canalize the Warsaw Pact attack. In combination, the two obstacles inhibit east-west movement because they have been either built up or dug in to provide a level surface for transportation.

The heavily wooded terrain between the IGB and Uelzen also would restrict the number of forces the Warsaw Pact could bring to bear at any one time up to the area immediately east of Uelzen. At that point, additional forces could be swung north, expanding the breadth of the attack.

If the Soviets can get the Dutch corps to commit its reserves in the south to hold Uelzen and prevent a breach of the Elbe-Seiten Canal line to the north of the city, they may find the Soviets attempting to outflank this defensive line from the north. In any such attempt the Soviets probably would mount heliborne assaults

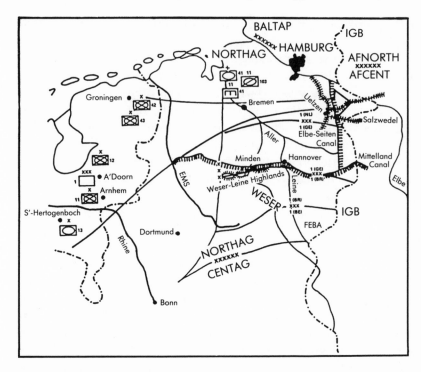

Figure 8.5 Dutch Corps Sector in NATO'S Northern Army Group

to seize the bridges that cross the Elbe River in FRG territory. By seizing the bridges in Hamburg, the Soviets could use the city as cover against German or Dutch counterattack.

If the Soviets should fail to cross the Elbe from the north, they probably would find it difficult to defeat the Dutch corps if it has deployed completely. Although the Dutch lack antitank helicopters, they have over nine hundred main battle tanks made up of Leopard IIs and the modernized Leopard I.[16] The most glaring vulnerability of the Dutch corps, aside from the fact that only a single reinforced brigade is deployed in Germany in peacetime, is its deficiency in ground-based air defense.[17]

Ideally, from their point of view, the Soviets would hope to get to the IGB defensive positions of the Dutch corps before the

Dutch were able to move the 350 kilometers from the Netherlands.[18]
If that deployment can be delayed or interrupted, the Soviets would
have the opportunity to shift the focus of defense in the Dutch corps
sector to the Aller-Leine and Weser River lines. Holding the Weser
and Leine rivers as far south as the Minden Gap and Hannover would
threaten operational-tactical scale encirclement of the I German
corps to the east of Hannover. Under such circumstances, to break
out of encirclement the German corps would have to conduct two
forced river crossings, assuming that the Germans even would be
disposed to abandon positions they might be defending successfully.
Furthermore, once the Soviets had crossed the Weser River in the
north, they probably would launch airborne or heliborne assaults on
the Ems River bridges and perhaps commit an army-level Operational Maneuver Group (OMG) to link up with these forces in an
attempt to cut off withdrawal of the Dutch corps. The Soviets would
not want NATO forces escaping to reestablish defensive lines that
would have to be penetrated again.

Soviet attempts to employ an OMG between the Leine and
Weser rivers or to seize the bridges across the Ems River could
encounter some problems. The terrain between the Leine and Weser
rivers, as well as that between the Weser and Ems rivers, is a dense
patchwork of moors. Although in many cases the German farmers
have drained the moors in order to farm the land, the underlying
peat is not strong enough to bear heavy traffic. Thus, even if the
Soviets were to manage to cross the Weser-Aller-Leine river line,
they might find it difficult to advance further west toward the Ems
Rivers or south between the Weser and Leine rivers if they were
forced to maneuver off the roads.

Now that the NORTHAG commander has divisions of his own
to commit to the battle, it is possible that the attempt by a first
echelon army to seize the Weser river line, or perhaps even the Aller-
Leine river line, could be defeated even if the Soviets beat the Dutch
to their positions along the IGB. The critical question in such an
instance, however, would be whether NATO had succeeded in preventing the Soviet air operation from severely reducing the early use
of the allied air forces in the ground battle. NATO's air power would
have to be available to be brought to bear decisively during these

first days of a war if the second echelon Warsaw Pact armies of the
first echelon fronts were to be prevented from concentrating in a
decisive manner at such a critical time in the battle.

As part of its plan to unravel the NATO coalition with early
operational success in the central region, the Soviets would call
upon the Danes, Dutch, and Belgians to capitulate and spare their
countries from inevitable ruin.[19] The Soviets are apparently convinced that the psychological and political impact of the loss of these
countries would be as significant as the military effect of destroying
them. If the alliance held firm, the Warsaw Pact would probably
attempt to seize Zeeland if it had not already been an immediate
objective with the initiation of hostilities.

The operational-strategic scale assault landing to seize Zeeland
would involve amphibious and airborne forces supported by naval
surface combatants as well as aircraft of the navy and the air forces.
In addition, the operation would be quickly reinforced by specially
trained motorized rifle troops that would be landed into the objective
area by air and sea.

All of the Danish islands are low and therefore do not present
any special difficulty for amphibious assault landings. Air cushion
landing craft and, in the future, wing-in-ground-effect craft would
probably be able to bring an assault force over any mine fields or
barriers to dry land. The Soviets would want to occupy Zeeland, but
forces defending some of the other islands might simply be isolated
by destroying Danish ferry ports and bridges. The operational objective would be to seize control of the Baltic Straits, but the political
objective would be to force Denmark to withdraw from the war in
the hope that the Netherlands and Belgium would follow Denmark's
example.[20]

The southern front on the north German strategic direction
also could have two initial operational directions: one active defensive
axis directed at fixing the I German corps in its IGB positions and
one offensive axis directed at the British corps sector. The defensive
axis directed at the I German corps sector might well involve an
entire army in the attempt to actively engage the corps east of
Hannover.

If the Germans were to hang on tenaciously to their defensive position along the IGB, the central front could conduct its main attack farther south. With the Mittelland Canal protecting its northern flank from a counterstrike by I German corps, a Warsaw Pact army could attempt to complete encirclement of the German corps through the British corps. The Salzgitter Canal and some low forested hills constitute the only terrain features the British corps can exploit in defending the open country to the north of the Harz Mountains. Although the Harz Mountains certainly do not constitute impassable terrain, the main operational directions likely would be through the Goslar Corridor to the north and the Gottingen Gap to the south. Thus a second Warsaw Pact army, probably subordinate to the northern front of the south German strategic direction, could attempt to move against the flank and rear of the British corps by cutting across the Belgian corps sector from south of the Harz Mountains.

South German Strategic Direction

Any attack against the British corps from the south would be channelized to the northwest by valleys that could be closed by mines and covered by artillery fire. These valleys, however, open into areas or "bowls" suitable for large tank battles. Successful breakout by Warsaw Pact forces to the north from the Bockenem bowl would allow them access to the plains north of Hildesheim. Once they crossed the Leine River, WP forces could use the city of Hannover as a shield against counterattack by the I German corps. The terrain would make a WP breakout to the northwest from the Einbeck bowl more difficult than movement north from Bockenem, but the results of such success could be more decisive. Forces advancing north along the Weser River could attempt to prevent NATO rescue of the I German corps by linking up at the Minden Gap with other WP forces moving south along the Weser River through the Dutch corps sector.

Even if a WP attack from the south through the Belgian corps sector were to fail to break into the depths of the British corps' defenses, it could divert considerable British resources from the defenses to the north of the Harz Mountains and make it difficult for the British to hold the line created by the Salzgitter Canal and the wooded hills west of the IGB. In the event of a breakthrough in the British corps sector from either the east or the south, the Soviets would probably commit a corps-sized OMG against the Rhine River and, subsequently, Bonn.

If Warsaw Pact forces were able to reach Bonn and turn south, they could attack the Schwabische Alps from the northwest. Of course, they would have to seize the bridges across the Rhine along their ever-lengthening right (west) flank. This would be no small task in that there are some eighteen road and rail bridges across the Rhine River between Bonn and Mannheim. Seizing these bridges would help to secure the right flank on this operational direction, and would also make it easier for a second echelon army to obtain the subsequent objective, the French border.

By turning south to Stuttgart, the Soviets might also hope to influence French decisions about how to react to rapidly developing events. The Soviets would be seeking to gain time. If the French could be induced to delay launching a decisive operation to break into the building encirclement of forward NATO corps in CENTAG, the army group probably would have to attempt a breakout. Even if the breakout were successful, it would probably leave the highly defensible terrain along the IGB and the FRG-Czech border in control of Warsaw Pact forces.

The Soviets might also have a very sound operational rationale for turning south at Bonn. They would want to close the encirclement as quickly as possible. Doing so would cut off NATO forward forces from ground resupply and reinforcement. The Soviets would expect that the presence of Warsaw Pact forces deep in NATO's rear areas would have a devastating effect on the morale of NATO forces engaged along the IGB.

To complete encirclement of the two German and two American corps in Central Army Group (CENTAG) the Warsaw Pact would have to attack these four corps along the IGB and the FRG-Czech border with enough force to hold them and, if possible, to induce them to commit reserves and to counterattack. The Warsaw Pact would then have to extend the southern arm of encirclement through northwestern Austria and the thinly deployed II German corps in

Bavaria. Advancing WP forces would not need to occupy Austria and probably would try to avoid fighting the Austrians. The Soviets probably would try to deal with Austria in one of two ways: (1) seize Vienna and hope to successfully preempt Austrian resistance; or (2) attempt to hold Vienna at risk in the south in the hope of obtaining freedom of passage in the northwest. Regardless of the option selected by the Soviets, the key operational components of the incursion through Austria would probably involve a heliborne assault on Linz and its airport and an airborne assault on Salzburg and its airport.

A heliborne assault on Linz would be directed to seize the shortest road and rail route from Warsaw Pact territory across the Danube River. The Linz airport is bordered on the north, east, and west by wide-open terrain and to the south by a military air base. An advance team of Soviet *spetznaz* forces probably would be inserted at night to suppress or delay initial Austrian reaction to the subsequent dawn attack by a heliborne assault force. Once the airfield had been secured the Soviets probably would need to air land additional forces before they would feel confident enough to move into the city to secure its critical road and rail bridges across the Danube River. The terrain on the axis toward Linz is rolling, but not difficult, and a new superhighway is being built from Linz toward the border with Czechoslovakia. The ground attack on Linz from Czechoslovak territory probably would be carried out by approximately a division-sized force moving on three routes of advance. If they were able to initiate their ground attack from the Czech border at the same time they were to launch the assault on the airfield, the Soviets probably would expect to be able to reach Linz to link up with the air assault force within twenty-four hours.

Twenty-four hours after the heliborne assault against the Linz airport, the Soviets probably would attempt an airborne assault on the airfield at Salzburg. A battalion assault to secure the airfield would have to be followed by air landing of larger elements to assist in seizing the city itself. In the course of the airborne assault on Salzburg airport, the Soviets probably would have committed an OMG to link up with the airborne force at Salzburg.

Unless major cities like Munich were turned into fortifications, the best terrain on which to deploy forces to stop the Warsaw Pact advance from the south would be the Schwabische Alps. If NATO should dig a division into the Schwabische Alps, it might become the anvil upon which NATO forward forces, or even Swiss forces, could strike the overextended attacker's forces on this operational direction.

The Soviets would hope to avoid any early entanglement with the Swiss by initially staying well north of their frontier. The potential operational impact of active opposition by the Swiss armed forces is of some concern to the Soviets.[21] Soviet planners would have to be especially worried about the Swiss response to Warsaw Pact violation of Austrian neutrality. The Swiss air force, with almost three hundred combat aircraft,[22] could greatly complicate Warsaw Pact airborne action against Salzburg. Any WP response against Swiss airfields could bring the attackers into full conflict with the Swiss air force and antiaircraft command. This would occur at a very inopportune time for the Soviets since the theater air operation already would be placing heavy demands on WP air resources. The additional requirement to suppress the twelve most important military airfields could simply be beyond the Soviet capacity.[23]

The Soviets might be equally disturbed, however, by the possible role of the sizable Swiss tank force. By 1993 the Swiss army will have a total of 380 *Panzer* 87 Leopard 2 tanks for a total of 900 main battle tanks.[24] Beyond the tank modernization program, Swiss artillery is to be increased by an additional six armored howitzer battalions.[25]

More than one-quarter of Switzerland is covered by mountains and glaciers.[26] It is understandable, therefore, that the Swiss need a mountain army corps to man the mountain redoubt system integrating the three fortress systems of Sargans, St. Gotthard, and St. Maurice, and extensive barriers and fortifications.[27] The purpose of a tank force approaching in size that of the French army's is less clear. The Swiss defense planner argues that, "although most of the terrain is unsuitable for mechanized warfare, the Swiss defense system includes full provision for countering aggressive tank movements through the valleys."[28] Furthermore, because Switzerland is crisscrossed with rivers, the Swiss army is well equipped with the combat bridging necessary to conduct mobile warfare.

The Soviets understand all too well that the impressive capability reflected in the three field army corps (each consisting of two field divisions and a mechanized division)[29] that the Swiss have deployed in the lowlands (the Plateau) not only would support a defense in depth against a Warsaw Pact attempt to cut through Switzerland from southern Germany toward southern France. These mechanized forces also would enable the Swiss to conduct their defense on the terrain of southern Germany should they choose to do so. Switzerland's selection of the German Leopard 2 over the American M1 Abrams tank to upgrade its armored forces only can add to Soviet suspicions that the Swiss might be trying to facilitate use of the Bundeswehr's logistics support system should they decide to fight alongside the II German corps in Southern Bavaria.

IMPLICATIONS FOR EUROPEAN NEUTRALS

Perhaps the clearest implication to emerge from our study is that the defense policies of the European neutrals make a difference in terms of peace and security in Europe. The effect is especially clear in the area of crisis stability.

In the north the neutrals threaten to impose serious constraints on Soviet operational planning. The Soviets must consider the difficult Finnish terrain and the proven determination of Finnish forces to fight when threatened as well as the fairly large and quite capable Swedish armed forces. These neutral forces must add considerably to Soviet uncertainties about the success of prospective operations both in the Northwestern TSMA and along the northern flank of the Western TSMA. The very capable Swedish air force is an especially difficult problem for the Soviets. They must plan either to engage those aircraft preemptively, on the ground if possible, thereby insuring Swedish hostility, or try to avoid Swedish airspace and considerably increase the exposure of Warsaw Pact air forces to attack by NATO air and air defense forces. In either case, neutral Sweden could greatly complicate WP seizure of Denmark and the Baltic Straits and thereby, at least indirectly, inhibit WP achievement of control of the Dutch and Belgian North Sea ports.

In the south tank and mechanized forces would greatly complicate Soviet planning for large-scale encirclement operations against CENTAG. Even the remote possibility that NATO forces in CENTAG might link up with Swiss and Austrian forces in a European redoubt anchored in the Alps and resupplied through the Mediterranean could mean a protracted conflict in Europe. The Soviets almost certainly would be deterred by the possibility of such a struggle, with its accompanying threat of eventual nuclear use and the possibility that their empire in Eastern Europe might begin to unravel in the absence of a fairly rapid victory.

Assessing the role of neutrals in European security from the perspective of a Soviet operational planner helps to bring into sharper focus what many have known for some time. The first, and most obvious, is that the security of the Western democracies is interlinked. Less obvious is the extent to which the security of the NATO states is bolstered by the commitment of each neutral state to defend itself. If the neutrals in Europe have been longtime "security borrowers," they are most certainly security contributors today. The trends, furthermore, suggest that future peace and stability in Europe is likely to be increasingly dependent upon the security contributions of the European neutrals along the East-West frontier.

PART THREE

PERSPECTIVES

CHAPTER NINE

OPTIONS FOR NEUTRAL STATES

Philip Windsor

It is somewhat paradoxical to discuss the role and options of the neutral states in the context of the East-West confrontation in Europe. After all, neutrality has classically meant noninvolvement in the conflicts of great powers, and has been a useful device for the prevention of conflict in areas where spheres of influence intersected. That function has not been entirely eclipsed today; but even so it is now clear that neutral states cannot simply opt out of the tensions between the preparations for war and the attempts to preserve peace which characterize contemporary Europe. They are as much involved in the force field of East-West relations as the members of the opposing alliances.

In the early years of the cold war one might have expected the opposite to be the case. Alliances in competition with each other were now the primary constituents of the international order, not merely the temporary and subordinate expedients that they had once been. That should surely have meant that Europe's neutral states had virtually no role to play. But it very rapidly became clear, as three European states—Finland, Yugoslavia, Austria—escaped the threat of Soviet domination, and as others—Sweden, Switzerland—determined to provide for their own security without joining the Atlantic alliance, that the political significance of neutrality was growing rather than diminishing. That fact, however, also meant that the neutrals would have to judge their conduct in the light of how they interpreted relations between East and West, just as one day they would find that they were able to influence those relations. Far from being excluded from the East-West framework by their status, they formed part of it.

But the paradox of their position went further: for they did not have a position. Neutrals can be defined as neutral, but they are all neutral in different ways. A country like Yugoslavia might even challenge the appellation neutral, preferring to call itself nonaligned. But while in global terms it behaves as a nonaligned country, in European terms it behaves remarkably like some of the other neutral states. Ireland, on the other hand, is, globally speaking, a neutral rather than a nonaligned country, but finds full-scale neutrality difficult to sustain in Europe, if only because of its membership in the EEC. Further, it is possible to group the neutral states in different ways: the glacis of Finland and Sweden in the north, or the south-central bloc of Switzerland, Austria, and Yugoslavia. But while the two northern countries might well observe that their neutrality is interdependent, and maintain close relations over a wide range of issues, neither the interdependence nor the closeness is to be seen among the other three. Insofar as one can generalize, particularity and differentiation are the qualities of the neutral states. Yet, at the same time, they are sometimes expected, and sometimes attempt, to make a common contribution to European security, particularly within the context of a developing East-West dialogue. In a sense, they have to generalize in spite of their particularity.

Such initial paradoxes help to clarify the scope, not only of the role of the neutral states but also of the context within which they operate. Earlier chapters have shown the prospects as well as the difficulties of maintaining a position of armed neutrality for some of the states concerned. Other authors have also argued that the neutrals make a highly effective contribution to the security of the West and of Europe as a whole. But both sets of considerations make it clear that European security cannot be defined in purely military terms. Even to discuss the stability of the military confrontation implies a wider understanding of the political relations between the two alliance systems. In fact, the security of Europe lies in a whole network of relations between East and West: political, economic, technological, and, perhaps above all, cultural; and the neutrals have a significant part to play in promoting and fostering them. They also have plenty of problems to confront, some of which will be discussed later. But their contribution to the security of Europe must be judged in this wider context, not simply in the narrower terms of the military confrontation.

To consider the military context first: it is clear, in the words of Lothar Ruehl in chapter 6, that the presence of neutral states contributes a "compensatory stability" to the risks of instability which might arise within the military confrontation. Ruehl analyzed in some detail the case of the Kola peninsula, and the manner in which Finland and Sweden contribute to the security of Norway and NATO as a whole. There are other, similar cases. One obvious example lies in the geographical relationship between Yugoslavia and Italy. The north Italian plain does not begin at the political frontier between the two but at the Ljubljana Gap. It is the neutrality of Yugoslavia which saves NATO planners from having to devote as much energy and as many resources to the defense of northern Italy as they do now to that of West Germany. And NATO would find it impossible, given the existing social and economic constraints, to do both. In this way, the very existence of neutral states can contribute to the stability of the military confrontation, and also help to keep the costs down. On the whole, as earlier chapters have made clear, the presence of the neutral states works to the advantage of the Atlantic alliance in any calculus of what might occur in the event of hostilities.

There are, however, difficulties, certainly for the future. The coming together of a number of new military technologies presages a real revolution in warfare. The development of the electromagnetic gun could make the tank obsolete—and in doing so could help to ease NATO out of its overdependence on the early use of tactical nuclear weapons. But not everything is stabilizing. "Stealth" technology might resolve many of the problems of aircraft penetration, and from the point of view of a defensive alliance this might also ensure that the opponent would be unwilling to risk attack in a crisis. But it might also heighten the fears of the opponent—particularly one apprehensive about the dangers of preemption—and in turn impel him to seize airfields in neutral territory. Such considerations raise two kinds of problems, even though, in principle, one might argue that neutral states would have a strong interest in the conventionalization of potential warfare, and indeed in the conventionalization of deterrence at the operational level.

The first problem is simply that of expense. Can a country like Sweden, which maintains an excellent, modern air force (and one of the largest in the world), also go to the lengths required to modernize its land forces and maintain an adequate seaborne defense? It is obvious that Sweden is already feeling the strain. Short of a totally unexpected upswing in the economic performance of the Western world as a whole, such strains are likely to increase for any individual neutral state even more than for the members of NATO, which are to some extent buttressed by the possibilities of alliance rationalization.

But considerations of emerging technology raise questions beyond that of expense alone. In part, these focus on the problems of maintaining a balance between the requirements of peacetime deterrence and preparations for active defense in the event of war. They are only partially congruent; and the incongruities might become sharper as the revolution in warfare proceeds and as the strategic context changes. To take one example: battlefield communications, the computerization of target acquisition techniques, the increasing ability in general to integrate C3I programs into a single system—all these mean that military decision-making will, in the near future, come closer and closer to real time. Yet the closer one gets to real time, the harder it becomes to exercise political control. Political control, however, will be essential if the conditions of active defense are not to degenerate into those of escalation. The consequence is that while emerging technology might help to sustain an apparently integrated posture of deterrence-and-defense, the problems of relating the two will be just as hard, to say the least, in the future as they have been in the past. In that respect, the role of the neutral states in a changing strategic context will demand constant reevaluation; and one cannot simply assume that such a role will always be objectively on the side of the West.

In turn, the forthcoming changes in the strategic context mean that new problems could arise. Their central concern might well be a consequence of a future stabilization of the military confrontation between East and West. That is now perfectly foreseeable in both technological and political terms. Some of the technological transformations have already been indicated here or addressed elsewhere in this book. Politically speaking, the possibilities evoked by Gorbachev in his attempt to break through the deadlock of the Vienna talks on mutual and balanced force reductions, or the agreements actually reached at Stockholm on confidence-building measures, indicate that the political will to achieve stability might be mustering strength. But the experience of Stockholm indicates that there could be new conflicts of interest between an alliance like NATO and the neutral states—particularly if the latter are trying to act in concert. In one sense it was a great achievement that in 1986 the participants in the Stockholm conference were able to stop the clocks and produce an agreement by a putative September 19 deadline, even though that agreement was more limited than had been hoped by the participating Western states. On the other hand, one of the reasons for stopping the clocks, quite apart from underlying distrust between the two protagonists, was a degree of tension, if not an outright conflict of interests, between the members of NATO and the neutral states in seeking to define what the criteria of stability and confidence might be.

Such considerations lead from the narrower context of the military confrontation to the wider, and more generally understood, one of European security. Conflicts of approach or interest in that wider context mean that the perception of the role of neutral states by the members of either alliance will continue to be important. In the case of the Western powers this role has already been acknowledged to a degree. Proposals for aerial surveillance, by the forces of neutral states, of military movements on either side of the confrontation in Europe indicate that the members of NATO regard the potential contribution that neutral countries might make to the establishment of a common understanding of security as both valuable and reliable. At the same time, however, one can still hear echoes in both Western Europe and the United States of an earlier view,

immortalized by John Foster Dulles when he declared that "neutralism is immoral." (He was speaking, of course, of neutralism not
neutrality, but when he made this remark in 1955, his strictures
were pretty undifferentiated. Later he was to modify them.) Today,
the accusations are grounded less in moral principle than in a sense
of resentment at the notion that the neutral states are getting a free
ride on the back of Western defense efforts. Earlier chapters have
made it amply clear that such sentiments are unjustified, both in
principle and in fact; but a better understanding of the fundamental
realities that have already been addressed here is still needed.

The neutral states make a positive contribution to European
security and incidentally save the Western powers both anxiety and
expense. Moreover, given an adequate and reasonably sympathetic
consideration of their potential role, they might contribute further,
especially in terms of crisis stability. At present their role in crisis
stability is largely passive: the compensatory stability indicated by
Lothar Ruehl arises from their sheer existence; the formidable obstacles that they present to would-be aggressors, as described by John
Hines and Phillip Petersen, obviously depend on more than pure
passivity. Yet location and terrain play at least as important a part
as active preparations for defense. The picture could change, though,
if the tentative establishment of a network of confidence-building
measures were to be extended. In such a case, neutral states would
have an increasing, and perhaps increasingly indispensable, part to
play. It might be that progress in arms control—depending, as it
is bound to if it ever comes about, not only on verification but
also upon the fostering of trust—will come to attach increasing
importance to confidence building. The fostering of trust, if a process
of that nature continues, will in turn depend on greater participation
by neutral countries. That much should be well understood by the
members of the Western alliance, particularly in terms of crisis
stability.

But there is a snag. The reaction of the Soviet delegation at the
Stockholm conference to the Western proposals for aerial surveillance
by the neutral states showed that, in the Soviet perception, there is
very little room for what one might call "more neutrality." That
might sound like an odd proposition. Wooing certain of the member

states of NATO into a condition of neutralism (again, not the same thing as neutrality) has long been a part of Soviet propaganda and diplomacy. On the other hand, the available evidence suggests that Moscow would be most reluctant to accord any specific or accepted role to the concerted action of the neutral states. "More neutrality" would in that sense be quite unacceptable, as was shown by Soviet reactions at Stockholm, and before that at Helsinki, Geneva, Belgrade, and Madrid. (Madrid has of course opted since that time for less neutrality.) There are possibly two reasons.

One is that in spite of all the vapid talk in many Western circles about Soviet ambitions to "Finlandize" Western Europe, the real danger in Soviet eyes is that Western Europe might gradually succeed in Finlandizing its Eastern counterpart. Such would at least provide a more accurate reflection of the realities of Finland's situation and behavior than the cheap usage which has been made of the country's name. That apart, the Soviet fear is bound to be that the example of the neutrals might, in different ways (Finland, Yugoslavia, Austria), yet again proffer a powerful temptation in Eastern Europe. One of the ironies of history is that Hungary, where a democratic uprising was crushed thirty years ago by Soviet tanks just after the country had declared a position of neutrality, has in the intervening years moved closer and closer to dependence on the Western economic system; whereas Yugoslavia, which had succeeded in establishing its neutral independence years before, is becoming increasingly interlinked with COMECON. The Soviet leadership may well argue that in terms of the correlation of forces it is not necessary to maintain rigid military structures after all in order to win the battle for allegiance, and that in the long term other factors might count for more. The true battle could then be thought of as one between economic structures rather than military structures. But for all this, the example of neutrality raises powerful specters; and concerted action by neutral states is still something to be feared.

Behind that, there is of course another logical argument. As Curt Gasteyger has pointed out in his chapter, the USSR can have no use for permanent neutrality so long as it still defines itself as the USSR. At the same time, however, it might assign different roles to neutral states in a changing historical process. What is essential here is the relationship between insuring the stability of the status quo and providing for stability in Eastern Europe. The problem is that the two are by no means coterminous—and certainly not in a Soviet perspective. Recent history is ambiguous on this point, but one could say that the greater the stability of the status quo in terms of East-West relations, the greater the instability in the Soviet domain in Europe. The East German rising followed the death of Stalin and came in the midst of tentative Soviet discussions about there being a new relationship with the West. The Hungarian revolt followed the Austrian State Treaty (which in itself provided an inspiration to follow the course of neutrality), Khrushchev's denunciation of Stalin, and the Polish attempt to throw off the Soviet yoke in the conduct of Poland's internal affairs. The Prague Spring came about at the high point of East-West détente, and was closely related to the development of a new pattern of intra-European relations. The Solidarity campaign in Poland, while superficially appearing at a time when East-West relations were entering the stage that has been characterized as the second cold war, had in fact put down its roots during the period when the two superpowers were seeking to codify the rules of their strategic relationship. In all these cases what looks like the promise of stability in the West can look very like the promise of disruption in the East. The great exception to such a litany is of course the first: Yugoslavia broke from the Soviet embrace at the height of the cold war. But the exception does not diminish the apparent threat. Precisely as a socialist state, but one which is not aligned with the USSR, and one which follows a different "path to socialism," Yugoslavia still appears to be viewed with great suspicion in Moscow. The Soviet embassy in Belgrade exercises a discreet but fairly continuous pressure on internal Yugoslav decision making, particularly in the matter of publications, and the response of the Yugoslav authorities is not always as robust as it typically was in the days of Tito. Diplomatically and economically, the Soviet Union has attempted to neutralize Yugoslavia's neutrality, so to speak.

In view of both the general circumstances in Eastern Europe and of the particular example of Yugoslavia, it is unlikely that the Soviet Union would be tolerant for long of any attempt by neutral states to concert their diplomacy in East-West relations. On occasion they have attempted to do so in the CSCE process; but such sporadic attempts are the best that can be hoped for. Proposals to involve them more continuously in monitoring the state of East-West relations by, for example, a more active role in confidence-building would most probably encounter stout Soviet resistance.

In general, then, the role of the neutral states in both maintaining and changing the network of relations which constitutes European security will remain limited. If the Stockholm experience suggested that the NATO countries would find it difficult to accept a concerted role for the neutral states (and some of the most difficult negotiations at Stockholm were among the NATO powers themselves as they tried to maintain their own consensus) the Soviet Union would find it virtually impossible.

Considerations of that order do not, however, suggest that, even in the wider context, the role of the neutral countries would be negligible or irrelevant. The interaction of their military capability with their political activity has itself provided an important form for relating the narrow context of military security in Europe with the wider context of cultivating more stable relationships in other fields. This very fact can, of course, create problems of its own. The EEC, for example, finds it difficult to elaborate a proper framework for cooperation with the neutral states, and there are plenty of outright conflicts of economic interest between the members of the community and the neutrals. In some particular instances the conflicts can become both more generalized and more sensitive. A case in point is that of Austria, free on its own to import many items on the COCOM list, but which has on occasion been stood in the dock by otherwise friendly Western powers because it was perceived as being too ready to trade such items with the Soviet Union. There are no easy choices here. Even if the accusations are justified, it is still true that while Austria's actions did not help the West in strictly military terms, they did nonetheless contribute to a stabilization of the overall relationship—if only by giving the Soviet Union an incentive to maintain it!

Clearly, there are ambiguities as well as difficulties in the role that the neutrals can play. The real point, though, is that neutrality is *evolving*, along with the European situation. Juridically speaking,

neutrality is static; but in political and military terms it is dynamic.
At the beginning of this chapter, it was suggested that alliances in
competition have become the main constituent elements of the
international order. But that situation is also liable to change.
Relations between the two superpowers do not simply contain the
relations between the states of Eastern and Western Europe. In
any case, the superpowers have themselves experienced so many
vicissitudes in their relationship in recent years that it would be
impossible for the behavior of European countries to be conditioned
purely by reference to what the superpowers were up to. At the same
time, it is quite clear that the conduct of relations in Europe is far
from autonomous—as Germans on both sides of the divide are, in
particular, well aware. For the foreseeable future there will continue
to be a West and an East. But, as (intermittently and sometimes
painfully) the two sides extend the range of their cooperation and
engage in more political and cultural exchanges, as well as forms of
economic and technological collaboration, so the role of the neutral
states is likely to grow in importance. As was suggested earlier, this
does not mean a growing coordination of policy on their part; and
indeed its dimensions cannot yet be forecast. The changes in strategic
context will affect their ability to maintain a well-armed neutrality—
not necessarily to its detriment but certainly in a manner which will
impose hard choices. Potential changes in the political context, both
of superpower relations and of those in Europe, will probably mean
that neutral states have a part to play in Europe that is more active
than that of acting as hosts at international conferences and generally
being helpful. If Europe—even Western Europe—can still expect
an economic and technological future, it will also be necessary to
associate at least some of the neutral states with enterprises managed
on a European scale. CERN is an indication of what to anticipate.

All this comes down to saying that the dynamics of neutrality
are likely to change and that much is unpredictable. What is clear
so far is that the existence and activities of the neutral states help,
on balance, to do much to safeguard the defense and security of the
Western alliance. And what is predictable for the future, insofar as
anything can be, is that their contribution to European security will
be of increasing importance.

NOTES

CHAPTER ONE

This essay attempts to set forth official Swedish security policy views. However, it must be understood that it is based on the author's interpretation and understanding of this policy. It is not a formal or an authoritative presentation of Swedish policy and does not necessarily represent the views of the Swedish government or of the Swedish Defense Research Establishment, nor, even necessarily, those of the author.

1. Krister Wahlbäck, *Den Svenska Neutralitetens rötter,* UD informerar, no. 3 (Stockholm, 1984); published in English as *The Roots of Swedish Neutrality* (Stockholm: Swedish Institute, 1986).

2. In addition to eight fighter and reconnaissance aircraft of old models, five bombers and twelve fighters of the latest models were delivered. Although not very impressive in numbers, the twelve fighters represented one-third of all Swedish operational fighters at the time. See Erik Norberg, *Flyg i beredskap, Det svenska flygvapnet i omvandling och uppbyggnad 1936–1942* (Stockholm: Allmanna Forlaget, 1971), 100.

3. This phrase can also be translated as "nonalignment in peacetime," "not aligned in peacetime," or even "free of alliances in peacetime"—the last a literal rendition of the Swedish. Although "nonaligned" is a less clumsy expression than the alternatives, it tends to be avoided because of its association with the Nonaligned Movement. I have used "nonallied" throughout this chapter.

4. The Swedish contribution to UNICYP was reduced mainly to a police force in 1987.

5. Little has been written on Swedish participation in UN peacekeeping activities. A short overview can be found in Gunnar Jarring, "Swedish Participation in UN Peacekeeping Operations," *Revue Internationale d'Histoire Militaire,* 1984, no. 57.

6. Heinz Vetschera, "Neutrality and Defense: Legal Theory and Military Practice in the European Neutrals' Defense Policies," *Defence Analysis,* 1985, no. 1:51–64.

7. John Ellis van Courtland Moon, "Chemical Weapons and Deterrence—The World War II Experience," *International Security* 8 (Spring 1984):3–35.

8. Alf W. Johansson and Torbjörn Norman, "The Swedish Policy of Neutrality

in a Historical Perspective," *Revue Internationale d'Histoire Militaire*, 1984, no. 57:69–94.

9. Concerning the Swiss staff talks, see, for example, Rudolf Bindschedler, Hans Rudolf Kurz, Wilhelm Carlgren, and Sten Carlsson, "Schlussbetrachtungen su den Problemen der Kleinstaatneutralität im Grossmachtkrieg," in *Schwedische und Schweizerische Neutralität im Zweiten Weltkrieg*, ed. Rudolf Bindschelder et al. (Basel: Helbing & Lichtenhahn, 1985), 434–35.

10. The data on Swedish defense expenditure were furnished by Bengt-Göran Bergstrand of the Swedish Defence Research Establishment.

11. A Norrland brigade is essentially an infantry brigade especially trained and equipped (including oversnow vehicles) for fighting in terrain and climate typical of northern Sweden.

12. As a result of the 1987 defense decision, the number of submarines will increase by two.

13. Figures 1.1, 1.2, and 1.4–1.7 are based on data from Kurt Törnquist, *OPINION 84–En opinionsundersökning hösten 1984* (Stockholm: National Psychological Defence Planning Committee, 1984); data for figure 1.3 were published in *Forsvarets forum*, 1983, no. 22:13. Subsequent data confirm the trends in these diagrams; see Göran Stütz, *OPINION 87* (Stockholm: National Psychological Defence Planning Committee, Report No. 146, December 1987).

14. Wilhelm M. Carlgren, *Swedish Foreign Policy during the Second World War* (London: Ernest Benn, 1977), 54–73.

15. Sandler was Social Democratic prime minister (1925–26) and minister for foreign affairs (1932–36 and 1936–39). He left the government after protesting Swedish nonintervention in the Finnish Winter War.

16. Carlgren, *Swedish Foreign Policy*, 114–21.

17. Johansson and Norman, "The Swedish Policy of Neutrality," 69–94.

18. One such change in Sweden's strategic environment which is the subject of much attention today is the growth of the Soviet Northern Fleet. Although it is not directed against the Nordic states, and has not been accompanied by an increase in Soviet land forces suitable for intervention in the region, this growth in maritime resources is perceived as having increased the strategic importance of the region. This expansion has, in fact, been going on continuously since World War II, and it has also resulted in a gradual reinforcement of land and air forces in the northern parts of Norway, Sweden, and Finland.

Similarly, the introduction of sea-launched cruise missiles on U.S. attack submarines and surface combatants in the Norwegian Sea has given rise to an increased attention in Sweden to means of countering low-level penetrations of Swedish airspace.

19. *Sweden's Security Policy: Entering the 90s. Report by the 1984 Defence Committee* (Swedish Official Reports Series, 1985), 23.

20. See, for example, pp. 18, 53.

21. This is the only instance where the results of the 1987 survey show a

significant deviation from the trends of the last decade. While still a clear minority, as many as 34 percent believed that the prospects of Sweden being able to stay out of a major European war are high or rather high, with 58 percent believing the opposite. At least part of the explanation for this result may be that the question was reformulated in the 1987 questionnaire.

22. This would be a "real" Clausewitzian war, as opposed to an "ideal" war. See Michael Howard's introduction to Karl von Clausewitz, *On War* (Princeton, N.J.: Princeton University Press, 1976).

23. *Sweden's Security Policy,* 54. One caveat: I have referred to this report as giving the official view. And, of course, it does give an official view, in the sense of a politically sanctioned public statement. Clearly this report is not a scientific text, but rather a highly political one. Still, it is interesting to compare the views on a European war expressed in this report with more alarmist statements sometimes associated with Swedish politicians.

CHAPTER TWO

1. For historical references, see, e.g., Max Jakobson, *Finnish Neutrality: A Study of Finnish Foreign Policy Since the Second World War* (London, 1986); D. G. Kirby, *Finland in the Twentieth Century: A History and Interpretation* (London, 1979); and Rene Nyberg, *Security Dilemmas in Scandinavia: Evaporated Nuclear Options and Indigenous Conventional Capabilities,* Cornell University Peace Studies Program, Occasional Paper no. 17 (Ithaca, N.Y.: June 1983).

2. The phrase was used by Anders Thunborg, Swedish minister of defence, at a Salzburg seminar in September 1984. It was published as "Sweden's Security Policy," *Österreichisches Jahrbuch für Internationale Politik 1984* (Vienna), 101.

3. There is no dearth of literature on the military situation in the far North. See, e.g., "Styrkeforholdet in Nord-Europa," *Militaerbalansen 1984–85* (Oslo), published in English in *The Military Balance 1984–85* (London: IISS, 1984), 104–26; Johan Jörgen Holst, "Norwegian Security Policy: The Strategic Dimension," in *Deterrence and Defence in the North,* eds. J. J. Holst, K. Hunt, and A.C. Sjaastad (Oslo, 1985), 93–133; Bo Huldt, "The Strategic North," *The Washington quarterly* (Summer) 1985:99–109; and Sverre Lodgaard, "The Big Powers and Nordic Security," in *Security in the North,* eds. Bo Huldt and Atis Lejins, Swedish Institute of International Affairs, Conference Paper no. 5 (Stockholm: 1984), 19–39.

4. One of the most cogent arguments to that effect is Gustav Hägglund, "Strategisk utveckling i Norden," *Flådestrategier og nordisk sikkerhedspolitik,* Bind 2 (Kopenhavn: Det sikkerheds og nedrustningspolitiske udvalg, 1986), 41–54.

5. For an argument in favor of the new strategy, see Admiral James D. Watkins, "The Maritime Strategy," *The Maritime Strategy* (supplement to *U.S. Naval*

Institute Proceedings), January 1986. For a critical view, see John J. Mearsheimer, "A Strategic Misstep: The Maritime Strategy and Deterrence in Europe," *International Security* (Fall) 1986:3–57.

6. A more comprehensive view of Finland's security position than can be given here is provided in *Report of the Third Parliamentary Defence Committee* (Helsinki, 1981).

7. Ibid., 20–22.

8. There are two treaties defining the status of the Åland Islands: Ahvenanmaan saarten linnoittamattomuutta ja neutralisoimista koskeva sopimus, 28 January 1922; and Suomen ja Sosialististen Neuvostotasavaltain Liiton valilla solmittu sopimus Ahvenanmaan saarista, 22 October 1940.

9. The English text of the treaty can be found in Roy Allison, *Finland's Relations with the Soviet Union 1944–84* (London, 1985).

10. For a full discussion of the process of getting the interpretation accepted in the missile case, see Max Jakobson, *Veteen piirretty viiva* (Keuruu, 1980), esp. 217–20.

11. For an English translation of the text, see Allison, *Finland's Relations with the Soviet Union,* Appendix 2.

12. This is stated in Article 1. The following article maintains that "The High Contracting Parties shall confer with each other if it is established that the threat of an armed attack as described in Article 1 is present."

13. See Klaus Törnudd, *Sanat ja teot* (Helsinki: Kirjayhtymä, 1982), pp. 33–35.

14. The last two reports have been translated into English and are available in the Committee Report series published by the Government Publishing Center.

15. *Parlamentaarisen puolustustoimikunnan lausunto puolustusministeriön hallinnonalan toiminta- ja taloussuunnitelmasta vuosille 1987–1991* (Helsinki, 1986).

16. *Report of the Third Parliamentary Defence Committee* (Helsinki, 1981), 41.

17. Ibid.

18. Ibid., 31.

19. It should be mentioned that all of Finland's fighter-interceptors have been purchased since 1975.

20. The point was strongly emphasized in the Defence Commission's report of April 1986. See pp. 24–25.

21. *Report of the Third Parliamentary Defence Committee,* 35.

22. It should also be pointed out that unlike Sweden, Finland has not been a target of submarine intrusions. Since 1980 Finnish territorial waters have been violated only three times.

23. *Report of the Third Parliamentary Defence Committee,* appendix, 76.

24. See Pauli Järvenpää, "Puolustusmenot ja niiden vaihtelu eri maissa," in *Suomolaiset ja turvallisuuspolitii ka* (Helsinki, 1986), 67–77.

25. Ibid.

26. Speech by Veikko Pihlajamaki, Finnish Minister of Defense, at the opening of the 101 Course on National Defense, 17 November 1986.

27. *Report of the Third Parliamentary Defence Committee,* 56–58.
28. See note 26.
29. Ibid.
30. For a good overview of Finland's domestic production, see Jacques Lenaerts, "Finland's Defense Industry," *International defense review,* no. 3 (1984):277–81.
31. Järvenpää, "Puolustusmenot ja niiden vaihtelu eri maissa," 73.
32. *Parlamentaarisen puolustustoimikunnan lausunto puolustusministeriön hallinnonalan toiminta- ja talaoussuunnitelmasta vuosille 1987–91,* 4.
33. Ibid., 18.
34. Ibid.
35. A speech by President Urho Kekkonen at the Swedish Institute of International Affairs, Stockholm, 8 May 1978. Reprinted in English in *The Yearbook of Finnish Foreign Policy 1978,* 64–66.
36. Pros and cons of the nuclear-free zone are discussed, for example, by Sverre Lodgaard and Per Berg, "Nordic Initiatives for a Nuclear Weaponfree Zone in Europe," *SIPRI Yearbook 1982* (London, 1982), 75–93.
37. See, e.g., Barry R. Posen, "Inadvertent Nuclear War: Escalation and NATO's Northern Flank," *International Security* 7 (1982):28–54. See also Desmond Ball, "Nuclear War at Sea," *International Security* 10 (Winter 1985–86).
38. Posen, "Inadvertent Nuclear War."
39. Thomas C. Schelling, *Arms and Influence* (New Haven, Conn.: Yale University Press, 1966).
40. See Ball, "Nuclear War at Sea."
41. For example, see Johan Jörgen Holst, "The Challenge from Nuclear Weapons and Nuclear Weapon-Free Zones," *Bulletin of Peace Proposals,* 1981, no. 3.
42. James Goodby, "Security for Europe," *NATO Review,* June 1984, no. 3:9.
43. The research has been organized and led by the Institute of Radiochemistry at Helsinki University under Professor Jorma K. Miettinen.
44. See, for example, Lynn R. Sykes and Jack F. Evernden, "The Verification of a Comprehensive Nuclear Test Ban," *Scientific American,* October 1982, 47–55; and Farooq Hussolin, "The Future of Arms Control: Part IV, The Impact of Weapons Test Restrictions," *Adelphi Papers,* no. 165 (London: International Institute for Strategic Studies, Spring 1981).

CHAPTER THREE

1. See, e.g., William F. Bader, *Austria Between East and West, 1945–1955* (Stanford: Stanford University Press, 1966); Gerald Stourzh, *Kleine Geschichte des österreichischen Staatsvertrages* (Graz, 1975); Manfried Rauchensteiner, *Der Sonderfall* (Graz, 1979).
2. Many parts of the treaties are almost identical, as for example, the arms limitation clauses.

3. NATO had been founded in reaction to the establishment of Communist proxy governments in Eastern Europe by the Soviet Union. See Vojtech Mastny, "Europe in U.S.-USSR Relations, A Topical Legacy," *Problems of Communism* 37 (January/February 1988):16–29. After 1950, when Austria had successfully resisted a communist attempt to usurp power, the West expected Austria to join NATO after concluding the state treaty and initiating a paramilitary training program for police forces as a nucleus for future Austrian armed forces. See also Rauchensteiner, *Der Sonderfall*, 305–310.

4. See also Vojtech Mastny, "Kremlin Politics and the Austrian Settlement," *Problems of Communism* 31:37–51.

5. See also Heinz Vetschera and James Rocca, "Österreich in den Ost-West-Beziehungen, Parts I and II," *Österreichische Militärische Zeitschrift (ÖMZ)* 23 (1985):298–305.

6. See also Franz Freistetter, "Die strategische Lage Österreichs," *Allgemeine Schweizerische Militärische Zeitschrift* 41, no. 5 (1983):237; Wolfgang Danspeckgruber, "The Defense of Austria," *International Defense Review* 17, no. 6 (1984): 721. Austria's neutrality thus drove a wedge between NATO's central sector in Germany and southern sector in Italy, which had been linked via the Tyrol corridor under French control during the Allied occupation of Tyrol.

7. See also Alfred Verdross, *The Permanent Neutrality of Austria* (Vienna: Verlag für Wissenschaft und Politik, 1978), 75.

8. The obligation is enshrined in Article 5 of the Fifth Convention of the Hague, 1907.

9. Austria's participation in peacekeeping operations includes a field hospital with ONUC (1960–63), a field hospital with UNFICYP (1964–73), one infantry battalion with UNFICYP (1973–present), one infantry battalion with UNEF II (1973–74), one infantry battalion with UNDOF (1974–), observers with UNTSO (1967–present), and participation in staff functions with all mentioned peacekeeping forces, including several force commanders. *Handbuch für Soldaten im Dienst der Vereinten Nationen,* Truppendienst series, vol. 29 (Vienna: Carl Ueberreuter, 1985), 19–43.

10. For the first and second period, see Konrad Ginther, *Neutralität und Neutralitätspolitik* (Vienna: Springer, 1975), particularly "Forschungen aus Staat und Recht," vol. 31. Although the second period lasted for eight years more, Ginther's analysis proved correct, noting a shift away from the traditional understanding of neutrality in Austria.

11. See, in detail, Rainer Eger, *Krisen an Österreichs Grenzen* (Vienna/Munich: Herold, 1981), 15–72 (Hungarian crisis) and 73–122 (Czech crisis).

12. A people's initiative based upon 100,000 signatures equals a parliamentary motion and has to be voted in Parliament: Article 41(2) of federal constitutional law.

According to the theory of nonviolent defense, this strategy could only defend the social structure but not the territory of the attacked state. See Gene Sharp, *National Security through Civilian-Based Defense* (Omaha: Association for

Transarmament Studies, 1985). For this reason it is unsuited for the defense needs of a neutral that has to defend territory: H. Vetschera, *Soziale Verteidigung, Ziviler Widerstand, Immerwährende Neutralität* (Vienna: Braumüller, 1978), 134–51.

13. For example, he had argued in 1955 that neither side had suffered a strategic disadvantage by withdrawing from Austria, ignoring the fact that NATO had lost the "Tyrol link" between Germany and Italy, whereas no similar disadvantage occurred on the Eastern side. Bruno Kreisky, "Die österreichische Alternative," *Forum* 2, no. 17 (1955):166–67.

14. When a people's initiative was launched in 1982 against excessive spending for this purpose, the government argued that "a conference centre brings more security than interceptors," which were at that time demanded by the opposition.

15. This is especially true of some activities in the Middle East, as for example favoring the PLO and establishing close contracts with Libya's Colonel Qaddafi. These activities were, however, opposed (58.2 percent) rather than supported (41.5 percent) by the Austrian foreign policy elite. See Hanspeter Neuhold, "Internationale Entwicklungen bis zur Jahrtausendwende aus der Sicht eines Teils der 'aussenpolitischen Elite' in Österreich," *Österreichische zeitschrift für Aussenpolitik* 20, no. 3 (1980):208–11.

16. Voting patterns in the United Nations and other empirical data show that Austria still correlated closest with small Western European nations, such as Denmark and the Netherlands. See Paul Luif, "Österreich zwischen den Blöcken; Bemerkungen zur Aussenpolitik des neutralen Österreich," *Österreichische zeitschrift für politikwissenschaft* 11, no. 2 (1980):209–20.

17. Bundeskanzleramt, *Landesverteideigungsplan* (Vienna, March 1985).

18. It was finally decided in 1985 to acquire 24 Saab-35 Draken interceptors from the Swedish air force to bridge the gap to the fourth jet generation due to arrive in the mid-1990s. Austria's air force has used Swedish aircraft ever since the early 1960s (Saab-J-29 Tunnan and Saab-105 fighter bombers and Saab-91 trainers). Despite some resistance by local politicians and by the Greens, the first Draken arrived in Austria in June 1988, and scheduled training has begun.

19. This refers to the central sector; see *The Military Balance 1987/88* (London: International Institute for Strategic Studies), 231–32.

20. This may be true both in execution of long-range interdiction, as demanded by the Follow-on-Forces Attack (FOFA) concept, and for revived ideas of "conventional strategic air warfare." See, e.g., Lt. Col. D. J. Alberts, "Deterrence in the 1980s: Part II; The Role of Conventional Air Power," *Adelphi papers*, no. 193 (London: International Institute for Strategic Studies, 1984), 30–46. It may be further the case if NATO would increasingly rely on conventionally tipped cruise missiles for long-range interdiction. See, for example, Rose E. Gottmoeller, "Land-Attack Cruise Missiles," *Adelphi Papers*, no. 226 (London: International Institute for Strategic Studies, Winter 1987–88), 13–30, 45–57.

21. See Hanspeter Neuhold and Heinz Vetschera, *Austria's Security Policy* (Geneva: United Nations Institute of Disarmament Research, 1984), 23–26; Danspeckgruber, "The Defense of Austria," 724–29; and *Landesverteidigungsplan*, 52–66.

22. To a certain extent, Austria has thus developed a "Follow-on-Forces Attack" (FOFA) concept of its own, relying on ground-based guerrilla type forces rather than on Emerging Technologies (ET).

23. A federal constitutional law of July 1, 1988 now defines the Austrian armed forces as "militia-type." The militia structure has thus been made mandatory.

24. The same term is also provided for COs, making Austria the only European country where COs do not serve longer terms than regular soldiers.

25. See, in detail, Ernest König, "Die österreichischen Streitkräfte: Kontinuierliche Entwicklung und drohende Krise," *ÖMZ* 26, no. 4 (1988):297–306. The revised structure introduced "mobile battalions" in order to fill the gaps left by the inadequate development of territorial forces, .which was perceived by some as reverting to mobile warfare instead of area defense. However, it appears to be a stopgap measure rather than a change in doctrine.

26. See Heinz Vetschera, *Die Rüstungsbeschränkungen des österreichischen Staatsvertrages aus rechtlicher, politischer und militärischer Sicht* (Vienna: Gesellschaft zur Förderung politischer Grundlagenforschung, no. 5, 1985).

27. The peace treaties of 10 February 1947, with Italy, Hungary, Romania, Bulgaria, and Finland.

28. *Foreign Relations of the United States, 1945* (vol. II), 135–36.

29. Ibid.; Based on the United Kingdom's first draft against "long-range weapons" of a range of more than 30 kilometers or 20 miles for the Italian treaty.

30. Misinterpretations occurred not only on the Eastern but on the Western side; see the completely distorted presentation of the problem in "Austria May Purchase Soviet Missiles to Secure Revision of Neutrality Pact," *Washington Post*, 2 August 1984.

31. See its chapter on "civil defense," para. 2.1 (civilian protection), 104–17.

32. Article 58 (c) obliges the belligerents to take the necessary measures to protect the civilian population against the dangers resulting from military operations. As a preventive measure, it would also require preparations for the protection of the civilian population in areas where military operations would be likely.

33. See, for example, Report of the European Security Study Group (ESECS), *Strengthening Conventional Deterrence in Europe: Proposals for the 1980s* (London: Macmillan, 1983); Andrew J. Pierre, ed., *The Conventional Defense of Europe* (New York: Council on Foreign Relations, 1986).

34. Ibid.

35. Examples would be the U.S. Army's Air-Land Battle Doctrine (FM 100-5) or SHAPE's Follow-on-Forces Attack (FOFA) concept. See also Boyd D. Sutton, John R. Landry, Malcolm B. Armstrong, Howell M. Estes III, and Wesley K. Clark, "New Directions in Conventional Defence?" *Survival* 26, no. 2 (1984):50–78.

36. Called Special Operations Forces and others in the United States, and *SPETS-NAZ* in the Soviet Union; see also John M. Collins, *Green Berets, Seals and SPETSNAZ: U.S. and Soviet Military Operations* (Washington, D.C., and New York: Pergamon/Brassey, 1987).

37. Heinz Vetschera, "Recent Trends in European Terrorism," *Defence Analysis* 1, no. 4 (1985):286–89.

38. The official term is "Mutual Reduction of Forces and Armaments and Associated Measures in Europe (MURFAAMCE)," although MBFR is still better known to the public.

39. See the proposal for a nuclear weapons-free corridor in central Europe by the Palme Commission, *Common Security* (London: PAN, 1982), 146–47.

40. See above, footnote 22 Austria's strategic concept.

41. It would also complicate Austria's situation politically. If NATO were incapable of even limited counteroffensive operations, the only reason to maintain Austria's defense capabilities would be the WP's remaining offensive capabilities. Austria's armed neutrality would then appear as a unilateral advantage for NATO and might provoke pressure by the East on Austria to disarm, in turn favoring the East over the West.

42. Heinz Vetschera, *Confidence Building Measures (CBMs) and European Security* (Vienna: Institute for Strategic Research/National Defence Academy, 1986), 53–73. See also Otmar Höll, "The CSCE Process: Basic Facts," *CSCE: N+N Perspectives,* ed. Hanspeter Neuhold (Vienna: Austrian Institute for International Affairs, Laxenburg Papers no. 8, 1987), 9–21.

43. The same is also true for stable relations between the alliances and other neutral or nonaligned European states, especially in Austria's vicinity.

44. It would also relieve Austria from the burden of an increasing number of political refugees from Eastern Europe who leave their home countries for political reasons (suppression of human rights).

45. They derive from the customary law of permanent neutrality.

46. There has been less reason to develop a specific "neutral identity" in other areas, as for example the third basket. Here, the Western orientation of the neutrals prevailed over the N+N identity, although the neutrals refrained from joining the West in confrontational approaches toward the human rights issues.

47. For example, Austria's air force consists mainly of Swedish and Swiss planes, Austria's air defense is basically Swiss-made. On the other hand, Austria has sold military vehicles to Switzerland.

CHAPTER FOUR

1. "Konzeption der Gesamtverteidigung," (Bern, 1973), 21.

2. Ibid., 4.

3. Hans Thalberg, "The European Neutrals and Regional Stability, in *The*

European Neutrals in International Affairs, eds. H. Neuhold and H. Thalberg (1984), 127.

4. V. Suvorov, *Inside the Soviet Army* (New York: Macmillan, 1982), 28.

5. From Jan Sejna, *We Will Bury You* (London: Sedgwick & Jackson, 1982), 121. "Under the Strategic Plan, there was no intention of respecting the neutrality of Switzerland. Despite its affirmation of non-alignment, we included its Army in our count of NATO forces. We considered Switzerland a bourgeois country and a fundamental part of the Capitalist system. The Plan stated that there was no chance of establishing Socialism there by peaceful means. Even the working class was 'aristocratic,' in the sense that it was dominated by skilled craftsmen who were even less promising material for the progressive movement than the middle class. However, until 1963 our military operational plans had recognized its neutrality, and that of Austria and Sweden. Then Marshal Malinovsky told us that this was a 'reactionary position.' 'In the forthcoming struggle between Capitalism and the proletariat,' he said, 'no one can be neutral. It would be a betrayal of the working class for any commander to respect Capitalist neutrality.' "

6. John McPhee, "A Reporter at Large," *New Yorker,* 31 October 1983, 64.

CHAPTER FIVE

1. Pierre Maurer, "Defence and Foreign Policy: Switzerland and Yugoslavia Compared," In *Yugoslavia's Security Dilemma: Armed Forces, National Defence and Foreign Policy,* eds. Marko Milivojevic, John B. Allcock, and Pierre Maurer (New York: Oxford University Press), 187.

2. Josip Broz Tito, *Govori i članci,* vol. 1 (Zagreb, 1959), 271.

3. *Službeni list FNRJ* 1945, no. 40:341.

4. *Isvestija,* 16 April 1945.

5. Savez Komunista Jugoslavije, *The Correspondence between the CC of the CPY and the CC of the CPSU* (Belgrade, 1948).

6. *Borba,* 9 September 1949.

7. Tito, *Govori i članci,* vol. 5, 21.

8. For a discussion of Yugoslavia in Western military planning during that period as well as Yugoslav-Western military contacts, see Beatrice Heuser, "Yugoslavia in Western Military Planning, 1948–1953," in Milivojević et al., pp. 126–163.

9. Quoted in Dennison Rusinow, *The Yugoslav Experiment, 1948–1974* (London: C. Hurst and Co., 1977), 44–45.

10. *Službeni list FNRJ* (1953, no. 15):153.

11. *Dodatak službenog lista FNRJ* (1954, no. 6):1.

12. *Pregled medjunarodnih ugovora i drugih akata od medjunarodnopravnog značaja za Jusoslviju* (Belgrade: Institut za medjunavodnu politiku i poivredu, Beograd, 1968), 976.

13. *Politika,* 28 December 1949.
14. *Politika,* 6 November 1950.
15. Alvin Z. Rubinstein, *Yugoslavia and the Nonaligned World* (Princeton, N.J.: Princeton University Press, 1970), 39.
16. Ibid., 18.
17. *Politika,* 7 October 1949.
18. United Nations, General Assembly, 5th Session, Resolution 378.
19. United Nations, General Assembly, 6th Session, Annexes, Document A/1/68.
20. In this context, see Rubinstein, *Yugoslavia,* and Leo Mates, *Nonalignment: Theory and Current Policy* (New York and Belgrade, 1972).
21. *Politika,* 13 February 1965.
22. Ibid.
23. *Politika,* 22 December 1954.
24. *Politika,* 23 December 1954.
25. Mates, *Nonalignment,* 75.
26. Odette Jankovitch and Karl P. Sauvant, eds., *The Third World Without Super Powers: The Collected Documents of the Non-aligned Countries,* vol. 1 (New York: Oceana Publications, 1978), p. 41.
27. Marshall R. Singer, *Weak States in a World of Powers: The Dynamics of International Relationships* (New York: The Free Press, 1972), 61–85.
28. In the context of Hungary, see Lars Nord, *Nonalignment and Socialism: Yugoslavia's Foreign Policy in Theory and Practice* (Stockholm: Raben & Szögren, 1973); in the context of both conflicts, see Irena Reuter-Hendrichs, *Yugoslawiens Osteuropapolitik in den Krisen des sowjetischen Hegemonialsystems: Eine Fallstudie zu den Entwicklungen in Ungarn/Polen (1956), der ČSSR (1968) und Polen (1980/81)* (Baden-Baden: Nomos Verlag, 1985).
29. *PlanEcon Report,* vol. 4, no. 10 (March 11, 1988), 17.
30. *Politika,* 16 May 1988 and *Danas* (Zagreb), 29 March 1988.
31. Yugoslavia is not an associated member of COMECON.
32. Pierre Maurer, in Milivojevic et al., p. 121.

CHAPTER SEVEN

1. Quoted from Urs Schwarz, *The Eye of the Hurricane: Switzerland in World War Two* (Boulder, Colo.: Westview Press, 1986), 154.
2. Schwarz, *Eye of the Hurricane,* 155.
3. Ulf Lindell and Stefan Persson, "The Paradox of Weak State Power," *Cooperation and Conflict* 21 (1986), quoted by Attila Agh, "The Strength of the Weak," unpublished paper, July 1988, 45.
4. See Raimo Vayrynen, "Neutrality and Non-Alignment: The Case of Finland," *The Non-Aligned World* 1 (July–September 1983).

CHAPTER EIGHT

1. Chief of the General Staff, Marshal of the Soviet Union S. F. Akhromeyev, "The New Political Thinking and Soviet Military Doctrine" (Speech to the Council on Foreign Relations, New York City, 11 July 1988).

2. As cited by Raymond L. Garthoff, "New Thinking in Soviet Military Doctrine," *Washington Quarterly* 11 (Summer 1988):134.

3. A very useful discussion of the terrain in NATO's area of forward defense—the Soviets refer to this as NATO's tactical zone of defense—can be found in Hugh Faringdon, *Confrontation: The Strategic Geography of NATO and the Warsaw Pact* (London: Routledge & Kegan Paul, 1986). This volume is weak on Soviet operational art and Warsaw Pact command and control, but its overall utility should make it indispensable for most Western military theorists.

4. *Finnish National Defense* (Helsinki: General Headquarters' Information Section, 1983), 12–13; and Lars B. Wallin, *Doctrines, Technology and Future War: A Swedish View* (Stockholm: National Defence Research Institute, 1979), 7. This latter book, which is a translation of the main text of the Swedish National Defence Research Institute (FOA) report A-10001-M3 (January 1979), represents the views of the seven defense scientists and defense planners asked by the Swedish Ministry of Defense to study the major powers' military policies as well as the interaction between them and developments in military technology.

5. V. Bestuzhev, "Combat Actions on the Sea," *Voyennaya mysl'* (Military thought), July 1971, no. 7, as translated in *Selected Readings from Military Thought*, Studies in Communist Affairs, vol. 5, part II, (Washington, D.C.: GPO, 1982), 104.

6. Ibid.

7. *Finnish National Defense,* 19.

8. Ibid.

9. *The Military Balance 1987–1988* (London: International Institute for Strategic Studies, 1987), 88.

10. See, for example, "This happens in the Swedish Air Force!" (Stockholm: Information Department of the Swedish Air Staff, June 1983).

11. V. Khomenskiy, "Anti-Submarine Warfare (According to Views of the NATO Command)," *Zarubezhnoye voyennoye obozrehiye* (Foreign Military Review), January 1984:79.

12. See Colonel Robert P. McQuail, "Khrushchev's Right Flank," *Military Review,* January 1964:7–16.

13. Heinz Vetschera, "Austria's Policy of Neutrality and European Security" (Paper presented at a conference on The Contribution of the Neutrals to Peace and Security in Europe, Geneva, 24–26 September 1986), 13.

14. Marshal of the Soviet Union N. V. Ogarkov, "Strategiya voyennaya" (Military Strategy), *Sovetskaya voyennaya entsiklopediya* (Soviet Military Encyclopedia), vol. 7 (Moscow: Voyenizdat, 1978), 564.

15. Lt. Colonel Jan Blumstein, "Frontal Aviation in an Air Operation," *Vojenska*

mysl (Czech version of Military Thought), August 1975, no. 8; Colonel Aleksander Musial, "Charakter i znaczenie operacji powietrznych we wspolczesnych dzialaniach wojennych" (The Character and the Importance of Air Operations in Modern Warfare), *Przeglad wojsk lotniczych i wojck obrony powietrznej kraju* (Polish Air Force and Air Defense Review), March 1982; and Phillip A. Petersen and John R. Clark, "Soviet Air and Antiair Operations," *Air University Review* (March–April 1985).

16. Interview with General G.L.J. Huyser, Chief of the Defense Staff for the Netherlands, *Journal of Defense and Diplomacy*, 4 (April 1986):16. "The primary task of the First Netherlands Army Corps, which is assigned to NATO's Northern Army Group Central Europe (NORTHAG), is to contribute to the defense of the North German plain. The other major task for the Netherlands Army is to defend the lines of communication leading through the Netherlands." From "Royal Kingdom of the Netherlands," *Journal of Defense and Diplomacy*, 4 (April 1986):35. General Huyser has explained that for the cost of forward deploying to Germany one additional brigade, it is possible to buy the transportation and railway system to move the whole Dutch corps on a more timely basis when it is needed. In addition, most of NATO's reinforcements are to come through Dutch and Belgian ports. A better transportation system will facilitate their movement as well. See Huyser, 16.

17. Huyser, 18.

18. Huyser, 16.

19. John J. Yurechko, "Command and Control for Coalitional Warfare: the Soviet Approach," *Signal*, December 1985.

20. See Christopher N. Donnelly and Phillip A. Petersen, "Soviet Strategists Target Denmark," *International Defense Review*, 1986, no. 8.

21. The two best recent examples of articles in the Soviet press highly critical of the Swiss armed forces' attitude towards NATO and attacking the wisdom of Swiss decisions to modernize its forces with NATO-compatible equipment appeared in V. Kuzentsov, "Training and 'Interpretations'," *Izvestiya*, 1 January 1985 and "In the Role of Potential Ally?" *Krasnaya Zvezda* (Red Star), 16 July 1985. The former article questions the training of Swiss pilots on a NATO air base on Sardinia. It implies that the Swiss air force is preparing to fight against Warsaw Pact aircraft and that the defense of a neutral country does not require sophisticated air training. The article also attacks the Swiss decision to purchase the Leopard II main battle tank. The latter article makes a special point of "the growing U.S.-Swiss military ties." We note with appreciation that our attention was drawn to these articles by Stephan Kux.

22. Ivo Sturzenegger, "The Swiss Air Force Today," *Soldat und Technik*, May 1988, no. 9.

23. Ibid.

24. Frank A. Seethaler, "Switzerland's Current Defense Posture," *International Defense Review*, June 1988, no. 6:627.

25. Brigitte Sauerwein, "The Swiss Military Procurement Program 1988," *International Defense Review,* June 1988, no. 6:631.
26. *The Defense Forces of Switzerland* (London: AQ & DJ Publications, 1983), 60.
27. Laurent F. Carrel, "Neutrality and National Defense: The Case of Switzerland" (Paper presented at a conference on The Contribution of the Neutrals to Peace and Security in Europe, Geneva, 24–26 September 1986), 23.
28. *The Defense Forces of Switzerland,* p. 39.
29. *The Army of a Small, Neutral Nation: Switzerland* (Berne: Army Training Group), 7.

INDEX

CONTRIBUTORS

Richard E. Bissell, who was Executive Editor of *The Washington Quarterly* at the time of this study, is now Assistant Administrator of the U.S. Agency for International Development. His professional specialty is economic security studies, in which neutral studies play a crucial role in Europe.

Curt Gasteyger is Director of the Programme for Strategic and International Studies and Professor at the Graduate Institute of International Studies, Geneva. He has written widely on the question of neutral states in world affairs, particularly in relation to post-World War II security issues in Europe.

Laurent F. Carrel is with the planning staff of the Ministry of Defense, Bern.

John G. Hines is a Soviet analyst with the Rand Corporation, Washington, D.C.

Pauli O. Järvenpää serves with the Ministry of Foreign Affairs, Helsinki.

Phillip A. Petersen is a senior Soviet analyst with the Department of Defense, Washington, D.C.

Jens Reuter is a researcher at the Sud-Ost-Institut, Munich.

Lothar Ruehl is State Secretary in the Ministry of Defense in the Federal Republic of Germany.

Heinz Vetschera is a defense analyst with the Landesverteidigungsakademie, Vienna.

Lars B. Wallin is Head of Soviet and East European Studies at the Swedish Defense Research Establishment, Stockholm.

Philip Windsor is a professor at the London School of Economics.